Earth First

Harnessing Space Technology to Solve Earth's Greatest Challenges

By Laura Lee

Lauxon Publishing, LLC

By leveraging advancements in space technology, humanity can address critical issues on Earth, from climate change and food security to energy sustainability and global connectivity.

Published by Lauxon Publishing, LLC

Author's Note:

Thank you for reading. As you make your way through these pages, I invite you to consider how space technology, once seen as something distant and abstract, can directly impact your life and the world around you. What we do in space today will shape the world of tomorrow, and I believe that through thoughtful, ethical applications, space exploration can help us address the most urgent challenges we face on Earth.

This book is not just an exploration of space technology—it is a blueprint for how we can use that technology to create a better future. It is a call for collaboration, responsibility, and innovation in the way we approach global problems. As we continue to reach for the stars, let us not forget that the solutions we seek may already be within our grasp, just waiting for us to use them in ways that benefit all of humanity.

I hope this work sparks a new understanding of space's potential to heal, protect, and empower us, and that it encourages you to take action—whether that be in the ways you think about the future, the policies you support, or the innovations you create. Together, we have the power to shape a future where technology, humanity, and the planet thrive in harmony.

Thank you for joining me on this journey. I look forward to seeing where it leads us.

Table of Contents:

Prologue:

In the vast expanse of space, humanity has always sought answers to questions beyond our reach. The stars have been our guiding lights, and the moon, our silent companion. But as we look outward, we must also look inward. Space exploration is not just about reaching new frontiers; it's about the very future of our planet and its inhabitants. What if space technology could unlock solutions to the problems we face right here on Earth?

This book is an exploration of how the technologies and innovations born from space exploration can serve as a lifeline for Earth. From revolutionizing our environmental management to improving our healthcare systems and rethinking energy usage, the vast potentials of space technology hold the key to a sustainable, prosperous future. But this journey is not just about technological advancement—it's about human cooperation, global responsibility, and the ethical imperative to ensure that space benefits everyone.

What follows is not merely a look into the future, but a call to action. Space technology offers answers, but it is up to us to ask the right questions, to create the policies, and to work collaboratively for the good of all. This is not a story of exploration for exploration's sake; it is a story of how the great adventure of space can illuminate the path toward a better, more sustainable Earth.

Part I: Foundations of Space Technology and Its Earthly Benefits

Chapter . 1

The Dawn of Space Technology

Understanding how the evolution of space technology can directly improve life on Earth.

The story of space technology begins with humanity's age-old fascination with the cosmos. Since ancient times, people have observed the stars, tracked the movements of planets, and imagined what lies beyond the sky. These observations, initially driven by curiosity and survival, gradually evolved into sophisticated knowledge systems that allowed for navigation and the measurement of time. As civilizations developed, the desire to reach beyond Earth intensified, inspiring inventors, engineers, and dreamers to design tools that could break the bonds of gravity. The first rockets, though rudimentary, marked the tangible start of humankind's journey into space. They were powered by a blend of scientific curiosity and the human instinct to explore, setting the stage for modern space exploration.

Early space technology emerged from military and scientific ambitions, creating a dual-purpose drive for innovation. Rockets developed for defense purposes were adapted to carry satellites and scientific instruments, revealing new possibilities for technology transfer. This shift demonstrated that tools designed for extreme environments could be applied to practical challenges on Earth. The launch of the first artificial satellite, a modest sphere of metal and circuitry, captured the imagination of billions and proved that humans could place instruments beyond our atmosphere. From that moment, the potential of space technology to influence everyday life became apparent. Scientists began to envision applications that ranged from communications to weather forecasting and environmental monitoring.

The evolution of space technology accelerated rapidly during the mid-twentieth century, fueled by international competition and scientific ambition. Governments poured resources into research programs, fostering innovation at a scale previously unimaginable. Engineers experimented with new fuels, materials, and designs, overcoming obstacles that had seemed insurmountable. Each successful launch, orbit, and mission expanded the realm of possibilities for technological application on Earth. Satellites provided real-time data about weather patterns, ocean currents, and atmospheric changes, offering solutions to problems once considered unsolvable. The momentum of this era laid the foundation for the diverse applications of space technology that continue to benefit humanity today.

Understanding the origins of space technology also requires examining the people behind the breakthroughs. Visionaries like Wernher von Braun, Sergei Korolev, and Katherine Johnson combined ingenuity with perseverance to push humanity beyond familiar limits. Their work demonstrated the power of interdisciplinary collaboration, merging mathematics, physics, and engineering to create tools capable of withstanding the harshness of space. These pioneers also showed that the impact of space exploration extends far beyond laboratories and launchpads. Innovations originally intended for space missions often find transformative applications in medicine, transportation, and environmental management. By tracing the evolution of these contributions, it becomes clear that space technology is not an abstract pursuit—it is a practical force shaping the modern world.

The dawn of space technology is inseparable from the promise of using knowledge gained beyond Earth to address terrestrial challenges. Satellites orbiting the planet collect vast amounts of data that inform agriculture, disaster preparedness, and climate science. Innovations in materials and propulsion enhance energy efficiency and manufacturing techniques on the ground. Communication systems developed for space provide critical connectivity to remote and underserved regions. Even our understanding of health and human physiology benefits from experiments conducted in microgravity. Each of these applications illustrates how the evolution of space technology has transformed from exploration for curiosity's sake into a strategic tool for improving life on Earth.

The rapid expansion of space technology was closely tied to advances in computing and electronics. Early satellites and spacecraft required intricate control systems capable of performing complex calculations in real time. Engineers developed miniaturized circuits and automated guidance systems that not only made space missions possible but also revolutionized industries on Earth. These technologies enabled faster communication, more precise navigation, and improved monitoring of critical systems in sectors ranging from aviation to shipping. The cross-pollination between space exploration and terrestrial technology illustrates the direct impact of orbital innovation on daily life. By pushing the boundaries of computation in extreme environments, space programs accelerated the growth of the digital era.

Material science also played a crucial role in the evolution of space technology. Engineers needed substances that could withstand extreme temperatures, radiation, and mechanical stress while remaining lightweight and durable. Innovations like reinforced alloys, heat-resistant ceramics, and composite materials were developed to meet these stringent requirements. These materials eventually found applications in construction, transportation, and consumer electronics, demonstrating how space-driven solutions address real-world problems. The rigorous testing required for spacecraft components set a new standard for reliability and efficiency in manufacturing. As a result, industries worldwide benefited from techniques and materials initially designed for space. This transfer of knowledge underscores the symbiotic relationship between exploration and everyday technological advancement.

Another defining aspect of early space technology was the ability to observe Earth from a new vantage point. Orbital satellites provided an unprecedented perspective, revealing patterns and phenomena that were previously invisible or poorly understood. Scientists could monitor weather systems, track environmental changes, and analyze land use on a global scale. This capability created new opportunities for managing natural resources, predicting disasters, and mitigating climate risks. The insights gained from these observations transformed agriculture, urban planning, and environmental policy. By leveraging the unique vantage point of space, humanity gained tools to address long-standing challenges on Earth with greater precision and foresight.

Public engagement played a significant role in sustaining the momentum of space technology. The excitement generated by televised rocket launches and mission successes inspired generations of students, engineers, and policymakers. Educational programs and outreach initiatives capitalized on this enthusiasm, promoting STEM education and career development. By connecting the public to the tangible benefits of space exploration, governments and organizations built support for continued investment. This societal backing proved critical for advancing technological capabilities and expanding international cooperation. Public interest, therefore, became not just a byproduct but an essential driver of innovation in the space sector.

The interplay between exploration and practical application became increasingly evident as missions progressed. Early satellite programs were designed to gather data, but their findings had immediate consequences for society. Weather satellites allowed farmers to plan planting schedules with unprecedented accuracy. Remote sensing provided information about water availability and soil health, helping mitigate drought impacts. Even simple technologies, like orbital communication relays, enabled more efficient coordination during emergencies. Each success reinforced the notion that space technology was not merely a scientific curiosity but a strategic resource for improving life on Earth. The lessons learned during these formative years laid the groundwork for future innovations that would expand the scope and scale of space-based solutions.

As space technology matured, international competition and collaboration both shaped its development. The Space Race between the United States and the Soviet Union accelerated innovation at an unprecedented pace. Governments invested heavily in research, training, and infrastructure, creating entire industries dedicated to aerospace development. At the same time, early collaborations, such as shared satellite data agreements, demonstrated the potential of cooperation to achieve common goals. These efforts highlighted that while space exploration can be driven by national pride, its applications can benefit the entire planet. By combining competition with collaboration, humanity advanced technology that directly improves life on Earth.

Space technology also transformed communication systems in profound ways. Before satellites, long-distance communication relied heavily on undersea cables and radio towers, which were vulnerable to failure and limited in reach. The introduction of orbiting satellites enabled instant transmission of data, voice, and video across continents. This breakthrough revolutionized industries such as banking, transportation, and media by creating a truly global network. Communities in remote or underserved areas gained access to information and connectivity that had previously been impossible. Moreover, these systems laid the foundation for today's internet infrastructure, proving that technology developed for space can redefine societal interaction. The ripple effects of satellite communication continue to influence education, commerce, and governance worldwide.

Scientific research conducted through early space programs offered insights that extended far beyond the missions themselves. Experiments in microgravity, for example, allowed researchers to study phenomena impossible to observe on Earth. These experiments contributed to a deeper understanding of physics, biology, and materials science. In turn, findings from these studies informed practical applications in medicine, engineering, and environmental management. Pharmaceuticals, water purification methods, and advanced materials often trace their roots to insights gained from space experiments. The continuous feedback loop between exploration and application underscores how investments in space technology yield tangible, life-enhancing results on Earth.

The ability to monitor Earth from space also reshaped environmental awareness. Satellites began to provide high-resolution images of glaciers, forests, and oceans, revealing rapid changes in ecosystems caused by human activity. Scientists could track deforestation, ice melt, and pollution trends with accuracy and timeliness never before possible. Policymakers and organizations gained powerful tools to guide interventions, conserve resources, and plan sustainable development strategies. The transparency offered by space-based monitoring encouraged accountability and informed public discourse. By observing the planet from orbit, humanity gained a clearer understanding of its environmental challenges and the means to address them effectively. This perspective shifted space technology from a scientific endeavor to a practical instrument for stewardship of the Earth.

The practical applications of space technology became increasingly apparent as satellite networks expanded. Weather forecasting transformed from a rudimentary and uncertain process into a precise, data-driven science. Meteorologists could track storms in real time, predict their paths, and issue warnings that saved countless lives. Similarly, agriculture benefited from satellite imagery, which allowed farmers to monitor soil health, plan irrigation, and optimize planting schedules. Governments and NGOs used remote sensing to manage natural resources more effectively and respond to crises like famine and drought. These examples illustrate that early investments in space technology had immediate, tangible benefits for everyday life.

Early satellite programs also enabled breakthroughs in navigation and logistics. Before GPS, travelers and transport systems relied on less accurate methods, often leading to inefficiencies and delays. Space-based navigation systems introduced a level of precision that transformed transportation, shipping, and even emergency response. Airlines could optimize flight paths, shipping companies could track cargo globally, and individuals gained reliable location information. Beyond convenience, these systems enhanced safety and efficiency, reducing fuel consumption and emissions. The integration of orbital technology into navigation exemplifies how space innovations can directly address pressing environmental and logistical challenges. Over time, GPS and related technologies became indispensable tools for commerce, security, and daily life worldwide.

Communication satellites expanded the reach of global media and information exchange. News could be transmitted instantaneously across continents, bridging cultural and geographic divides. Remote communities, previously isolated, gained access to educational programs, healthcare information, and emergency broadcasts. The technology also enabled telemedicine and real-time consultation, improving medical care in regions lacking specialists. Businesses could operate across borders with ease, creating economic opportunities that were impossible before space-based communication. The wide-ranging impact of these systems demonstrated that space technology could strengthen the social, economic, and political fabric of societies.

Space technology also contributed to disaster management and humanitarian aid. Satellites provided rapid assessment of areas affected by earthquakes, floods, and hurricanes. Relief agencies could determine the scale of damage, identify critical infrastructure needs, and coordinate the distribution of supplies efficiently. This capability reduced response times, minimized human suffering, and optimized resource allocation. In addition, predictive models based on satellite data helped communities prepare for potential disasters. By combining observation, modeling, and real-time communication, space technology transformed the way humanity responds to crises. These early successes underscored the potential of space innovation to safeguard lives and enhance resilience on Earth.

The dawn of space technology was marked by a profound realization: exploration and utility are inseparable. As missions grew more sophisticated, the line between scientific curiosity and practical application blurred. Technologies initially designed for extreme conditions in space found countless uses on Earth, from improving energy efficiency to enhancing communication networks. The challenges of operating beyond our atmosphere inspired solutions that made life safer, more productive, and more sustainable. By learning to leverage space technology, humanity discovered a powerful tool for addressing its most urgent problems. This foundational understanding would guide the development of more advanced systems capable of tackling complex, global challenges in the decades to come.

The legacy of early space technology continues to shape our world in profound ways. Technologies developed for satellites, spacecraft, and space missions have become integral to sectors ranging from agriculture and energy to communication and healthcare. Innovations such as remote sensing, GPS navigation, and orbital observation provide tools that allow governments, businesses, and communities to make data-driven decisions. These capabilities help address complex challenges like climate change, resource scarcity, and disaster management with unprecedented accuracy and efficiency. The evolution of space technology demonstrates that investment in exploration yields tangible benefits for life on Earth. It is a clear example of how human ingenuity can transform scientific curiosity into practical solutions that improve the planet.

Furthermore, space technology has inspired a global culture of innovation. The challenges of operating in extreme conditions encouraged problem-solving approaches that prioritize efficiency, sustainability, and resilience. These approaches have influenced industries ranging from manufacturing to healthcare, fostering cross-disciplinary innovations that directly enhance human well-being. By adopting the principles developed for space exploration, Earth-based systems have become more reliable, effective, and adaptable. The lessons learned in orbit inform everything from urban planning to environmental conservation, proving that the skills and knowledge gained in space have far-reaching applications. Humanity's ability to translate these lessons into everyday benefits marks one of the greatest achievements of the space era.

Education and workforce development have also benefited from the dawn of space technology. The excitement of space missions motivated generations to pursue careers in science, technology, engineering, and mathematics. Institutions developed programs focused on aerospace, satellite systems, and applied physics, cultivating a workforce capable of advancing innovation on Earth. Public interest in space exploration continues to fuel research funding, private investment, and international collaboration. By inspiring curiosity and promoting technical literacy, space technology has created a virtuous cycle of innovation and societal improvement. This human capital is as important as the physical technology itself, driving forward solutions to pressing global challenges.

The environmental impact of space technology demonstrates its transformative power for sustainability. Satellites allow precise monitoring of deforestation, water availability, and atmospheric changes, giving policymakers actionable data to protect ecosystems. Predictive models based on space-based observations support agriculture, energy planning, and disaster preparedness, minimizing waste and optimizing resource use. By providing a macro-level view of Earth, space technology enables targeted interventions that would otherwise be impossible. These applications highlight that space exploration is not detached from terrestrial concerns but deeply intertwined with the planet's health and humanity's survival. The dawn of space technology, therefore, is not just a historical moment—it is a continuing force for environmental stewardship.

Ultimately, the emergence of space technology illustrates the extraordinary potential of human ingenuity. From the first rockets to sophisticated satellite networks, each advancement has expanded the toolkit available to solve Earth's challenges. By observing, experimenting, and innovating beyond our atmosphere, humanity gained perspectives and solutions that fundamentally enhance life on our planet. The lessons of the early space era provide a roadmap for future exploration, showing that investments in science and technology can produce direct, measurable benefits. As we continue to expand our presence in space, the principles established during its dawn will guide efforts to harness technology for the common good. The dawn of space technology reminds us that the pursuit of the stars ultimately illuminates solutions for life on Earth.

Chapter . 2

From Rockets to Real-World Solutions

Exploring how launch systems and spacecraft innovations can drive practical applications for everyday life.

Rockets are often seen as symbols of exploration, but their development has produced innovations that profoundly impact life on Earth. The engineering challenges of building powerful, reliable launch systems required breakthroughs in materials, fuel technology, and propulsion design. These breakthroughs were later adapted for terrestrial applications, including energy-efficient engines, transportation systems, and industrial machinery. By pushing the limits of physics and engineering, rocket development created solutions that address everyday challenges in unexpected ways. Beyond their symbolic value, rockets became catalysts for technological progress that extends far beyond space missions. Understanding how these innovations transfer from orbit to Earth reveals the practical power of space technology.

The principles of rocket propulsion have influenced multiple industries by improving energy efficiency and performance. Engineers developed lightweight alloys and composites to withstand the extreme temperatures and pressures of launch, which later found applications in automotive and aerospace manufacturing. Fuel management systems designed for rockets inspired safer, more efficient energy storage solutions in industrial and commercial settings. Even fluid dynamics research from rocket testing improved the design of pipelines, ventilation systems, and turbines. These contributions demonstrate that solving extreme challenges in space drives advancements that enhance safety, efficiency, and sustainability on Earth. By examining the journey from rocket design to everyday innovation, the tangible benefits of space technology become clear.

Spacecraft systems also provide transferable technology for Earth-based applications. Life support systems designed to maintain controlled environments in space have influenced medical technology, including ventilators and advanced air filtration systems. Thermal regulation methods used in spacecraft inform climate control in buildings, reducing energy consumption while improving comfort. Even waste management solutions for long-duration missions have inspired approaches to recycling and resource efficiency in cities. These examples illustrate how innovations developed for extreme conditions in space often have practical applications for human health, infrastructure, and environmental management on Earth. The connection between spacecraft engineering and everyday problem-solving highlights the broader impact of space technology beyond exploration.

The iterative nature of rocket and spacecraft development has fueled a culture of continuous improvement and experimentation. Engineers learned to test, fail, and adapt quickly, creating robust systems capable of extreme reliability. This methodology has influenced sectors like manufacturing, logistics, and emergency response, where resilience and adaptability are critical. By studying how spacecraft are designed to endure harsh conditions, Earth-based engineers have developed more reliable machines, systems, and tools. These lessons also extend to software, robotics, and automated systems, reinforcing the link between space innovation and practical Earth applications. The knowledge gained from solving the challenges of rocket and spacecraft engineering serves as a blueprint for tackling complex, real-world problems.

Public perception and economic investment have further reinforced the real-world benefits of rocket technology. High-profile launches capture the imagination and inspire investment in research and development, creating a ripple effect that benefits multiple industries. Government programs and private companies alike have leveraged these investments to develop technologies that enhance communication, transportation, and energy management. The excitement surrounding space missions translates into broader support for technological innovation, fueling new solutions for everyday challenges. By connecting space exploration to tangible societal benefits, rockets serve as more than tools for reaching orbit—they become instruments for global improvement. This dual purpose of exploration and utility underscores the transformative power of space technology for life on Earth.

Rockets' influence extends far beyond their ability to reach space, shaping advancements in transportation technology. The high-performance engines designed for spacecraft inspired improvements in jet engines, hybrid propulsion systems, and fuel-efficient vehicles. Engineers adapted concepts like thrust vectoring and staged combustion to optimize performance and reliability in commercial applications. Even aerodynamic designs tested for launch vehicles informed the development of high-speed trains, wind turbines, and aircraft. These innovations demonstrate how solutions created for the extreme conditions of space travel can improve efficiency, safety, and sustainability on Earth. By studying rockets, industries gained insights into managing energy, reducing waste, and maximizing performance under challenging conditions.

Space launch infrastructure has also driven innovations with broad terrestrial applications. The precision required for launching rockets safely led to advancements in automated control systems, structural engineering, and robotics. Ground support technologies, including hydraulic systems and monitoring sensors, were adapted for industrial automation and construction projects. Launch pad designs influenced high-load infrastructure, such as bridges and skyscrapers, which require resilience against extreme stresses. These contributions show that space launch facilities are not only gateways to orbit but also laboratories for engineering solutions that benefit everyday life. The lessons learned from designing and operating these infrastructures continue to inform safe and efficient practices across multiple industries.

Fuel and energy innovations from rocketry have transformed energy management on Earth. Solid and liquid propellants developed for rockets required careful handling, efficient combustion, and energy optimization. These advancements led to better fuel systems for power plants, transportation, and industrial processes. The study of chemical propulsion also inspired cleaner and more efficient energy storage technologies, including batteries and fuel cells. Engineers applied principles of energy density and thermal management from rockets to improve the performance and safety of terrestrial systems. By pushing the limits of energy efficiency in space, rocket technology has indirectly enhanced the sustainability of multiple Earth-based industries. These improvements highlight the interconnectedness of space exploration and practical energy solutions.

Material innovations developed for rockets remain foundational to modern engineering. Lightweight composites, heat-resistant alloys, and carbon-based materials were originally designed to withstand extreme forces and temperatures during launch. Today, these materials are used in everything from medical devices to consumer electronics and renewable energy systems. The rigorous testing and quality standards required for rocket components also established best practices for durability and performance in civilian applications. By borrowing from aerospace materials science, industries have reduced failures, improved efficiency, and extended product lifespans. This transfer of knowledge demonstrates that space technologies often create ripple effects, enhancing the quality, safety, and sustainability of products used daily on Earth.

Rockets also provided a framework for integrating multiple complex systems into a cohesive whole. Each launch requires coordination of propulsion, navigation, communication, and life support systems working seamlessly together. Engineers applied these integration principles to sectors like manufacturing, transportation logistics, and emergency response systems. The ability to manage complex operations with precision improved efficiency, reduced errors, and increased safety across multiple industries. By learning from the orchestration required for rocket launches, Earth-based systems became more reliable and adaptable. This systems-level thinking, cultivated in the pursuit of space exploration, continues to guide technological innovation that addresses real-world challenges effectively.

Spacecraft design has provided numerous innovations that benefit life on Earth. Life-support systems, originally developed to sustain astronauts in extreme environments, have inspired advances in medical technology. Controlled oxygen and temperature systems contributed to improved ventilators, respiratory treatments, and hospital climate control. Similarly, water recycling and purification technologies designed for spacecraft informed methods for providing clean water in remote or disaster-affected regions. Thermal regulation systems from spacecraft have been adapted to reduce energy consumption in buildings and industrial facilities. These applications illustrate that solutions designed for life in orbit can solve pressing problems on the ground.

Satellite deployment has been a particularly transformative aspect of spacecraft technology. Communication satellites created reliable, global networks that allow for real-time information exchange, remote education, and telemedicine. Earth observation satellites provide critical data for agriculture, urban planning, and disaster preparedness. These platforms also enable climate monitoring, resource management, and environmental conservation at scales previously impossible. By translating the precision and reliability of space-based systems into terrestrial applications, satellites have enhanced efficiency, safety, and sustainability across industries. Their impact extends to nearly every facet of modern life, demonstrating the broad practical utility of spacecraft innovations.

Navigation and positioning systems derived from space technology have revolutionized transportation and logistics. GPS and similar systems enable precise tracking of vehicles, cargo, and individuals, improving safety and operational efficiency. Supply chains rely on satellite navigation to optimize routes, reduce fuel consumption, and minimize delivery times. Aviation and maritime industries depend on these technologies to coordinate movements, avoid hazards, and respond to emergencies. Even personal navigation devices and smartphone apps trace their origins to spacecraft-based systems. The integration of orbital navigation technology into everyday life highlights how innovations intended for space travel can enhance safety, productivity, and convenience on Earth.

Robotics and automation developed for spacecraft have created solutions for industrial and environmental challenges. Robotic arms, autonomous rovers, and automated maintenance systems were initially designed for remote operation in space. These technologies have since been adapted for manufacturing, medical surgery, and hazardous environment operations on Earth. They increase precision, reduce human exposure to danger, and improve efficiency across multiple sectors. The lessons learned from operating robots in orbit also inform artificial intelligence applications for monitoring, prediction, and task optimization. By transferring these capabilities from space missions to terrestrial uses, engineers have solved problems in sectors that require precision, reliability, and safety.

Power systems designed for spacecraft have led to advances in energy efficiency and sustainability. Solar panels, batteries, and energy management systems developed for space missions provide critical insights for renewable energy projects on Earth. Orbital solar systems require lightweight, durable materials and optimized energy conversion—principles that enhance terrestrial solar energy applications. Energy storage technologies, informed by spacecraft battery research, improve reliability and capacity for homes, businesses, and electric vehicles. The integration of these innovations contributes to reducing greenhouse gas emissions and promoting sustainable energy solutions. Rockets and spacecraft, therefore, not only take us to orbit but also provide technologies that make life on Earth more sustainable, resilient, and efficient.

The development of rockets and spacecraft has fostered a culture of innovation that extends far beyond aerospace engineering. Engineers and scientists learned to solve complex problems under extreme constraints, a mindset that has influenced industries ranging from healthcare to energy. The rigorous testing, iterative design, and risk management required for space missions instilled principles that improve efficiency, safety, and performance in terrestrial systems. Companies and research institutions have adopted these methodologies to enhance manufacturing, logistics, and infrastructure development. By translating space-based problem-solving strategies to Earth-bound challenges, innovations are implemented faster and more effectively. This culture of continuous improvement demonstrates how space technology provides not only tools but also a framework for solving real-world problems.

International collaboration in space programs has also advanced technological solutions for Earthly applications. Joint missions and shared satellite data have encouraged knowledge exchange across borders, improving research and innovation globally. Collaborative projects enable faster development of new technologies, such as advanced sensors, robotics, and renewable energy systems. The cooperative approach has shown that shared goals and combined expertise produce solutions that benefit multiple nations and communities. Lessons from these partnerships inform collaboration in disaster response, environmental management, and public health initiatives. Space exploration, therefore, not only generates technology but also models effective teamwork and cooperation on a global scale.

The commercial space sector has amplified the transfer of rocket and spacecraft innovations to everyday life. Private companies developing launch systems have introduced cost-effective solutions that were previously only achievable by large governments. Competition encourages efficiency, faster innovation, and the development of technologies with broad commercial applications. For example, reusable rockets reduce material waste and operational costs, inspiring similar approaches in manufacturing and logistics. Companies also invest in satellite constellations for communication, Earth observation, and navigation, directly supporting economic growth and public services. The commercialization of space demonstrates that innovation in rocketry has tangible, scalable benefits for society.

Education and workforce development have been transformed by advances in rockets and spacecraft. The excitement of space missions inspires students to pursue careers in science, technology, engineering, and mathematics, creating a pipeline of talent capable of solving Earthly problems. Universities and research centers offer programs in aerospace, robotics, and satellite systems that produce skilled engineers, data scientists, and innovators. Exposure to space technology fosters critical thinking, creativity, and interdisciplinary collaboration. These human capital investments ensure that the knowledge gained from space exploration continues to generate benefits in healthcare, energy, transportation, and environmental management. By inspiring and training the next generation, space technology strengthens society's capacity to address global challenges.

The long-term impact of rockets and spacecraft on life on Earth is both profound and far-reaching. Innovations developed to withstand the extreme conditions of space have produced solutions that improve safety, efficiency, and sustainability across industries. From energy management and transportation to healthcare and communication, technologies inspired by space missions have become integral to daily life. The lessons learned in propulsion, materials science, and systems integration continue to inform modern engineering practices, allowing humanity to solve complex problems more effectively. Rockets and spacecraft serve as a bridge between exploration and application, demonstrating that advancements in extreme environments can generate tangible benefits on Earth. By leveraging these technologies, society gains tools to address critical challenges and create a more resilient, sustainable future.

Beyond the direct technological contributions, the culture fostered by space exploration has lasting societal effects. The pursuit of challenging goals encourages creativity, critical thinking, and collaboration, qualities essential for innovation in any field. Industries on Earth have adopted the problem-solving approaches developed for space missions, including iterative design, risk assessment, and data-driven decision-making. These practices improve operational efficiency, reduce errors, and enhance safety across diverse sectors. By translating the lessons of space exploration into everyday problem-solving, humanity has expanded its capacity to address environmental, economic, and social challenges. The success of rocket and spacecraft innovation highlights the enduring connection between exploration and practical solutions for life on Earth.

Economic benefits also stem from the technologies and industries developed for rockets and spacecraft. Satellite-based communication and navigation systems have created entirely new markets and business opportunities, fostering global economic growth. Reusable launch systems reduce costs for commercial and scientific missions, allowing a broader range of stakeholders to access space technologies. These economic incentives accelerate the development of solutions that can improve healthcare, disaster management, and environmental monitoring on Earth. Investment in space technology thus produces a cycle of innovation that drives both societal and economic advancement. The interplay between economic growth and technological progress ensures that space exploration remains a powerful engine for improving life on our planet.

Rockets and spacecraft have also strengthened humanity's ability to respond to global challenges. Satellite networks provide real-time data on climate, natural disasters, and resource availability, allowing governments and organizations to make informed, timely decisions. Navigation systems improve transportation efficiency and disaster relief coordination, saving lives and reducing environmental impact. Life-support, energy, and recycling technologies from spacecraft inform solutions to health, sustainability, and resource management challenges. By adapting these technologies to terrestrial needs, society benefits from the precision, reliability, and innovation originally developed for space exploration. The ability to transform challenges into opportunities through technological adaptation demonstrates the lasting value of rockets and spacecraft.

Ultimately, rockets and spacecraft exemplify how human ingenuity in one domain can generate wide-ranging benefits for the world. Technologies developed for exploration of space have become indispensable tools for addressing Earthly problems, from energy sustainability to disaster response. The collaborative, innovative, and problem-solving approaches cultivated through space programs continue to inspire advancements across industries. By understanding and leveraging these lessons, humanity is equipped to confront complex challenges with creativity and precision. The journey from rockets to real-world solutions illustrates the transformative power of space technology, proving that the pursuit of the stars ultimately improves life on Earth. The legacy of these innovations lays a foundation for continued progress, guiding the next generation of space-based solutions to benefit humanity and the planet.

Chapter . 3

Satellites as Tools for Global Problem-Solving

Demonstrating how satellites provide critical data to tackle climate, health, and agricultural challenges.

Satellites have transformed the way humanity observes and interacts with Earth. Once merely instruments for communication and exploration, they now serve as essential tools for understanding global systems. From monitoring weather patterns to tracking deforestation, satellites provide data that informs decisions across multiple sectors. Governments, researchers, and organizations rely on this information to manage natural resources, plan infrastructure, and respond to crises. The ability to gather accurate, real-time information from orbit has redefined problem-solving strategies for environmental, health, and social challenges. By leveraging satellite technology, humanity can address complex issues with precision and foresight previously unimaginable.

Earth observation satellites play a critical role in climate monitoring and environmental management. Sensors onboard these satellites measure atmospheric temperatures, greenhouse gas concentrations, and ocean currents. They track ice cover, desertification, and changes in vegetation, providing essential insights into global warming and ecosystem health. This information enables scientists to model future climate scenarios and develop strategies to mitigate the effects of environmental change. Policymakers can use these data-driven insights to implement regulations and initiatives that protect natural resources. Satellites, therefore, serve as indispensable eyes in the sky, guiding decisions that directly affect the planet's sustainability.

In agriculture, satellites provide tools for improving food security and resource management. High-resolution imagery allows farmers to monitor crop health, detect stress, and optimize irrigation practices. Remote sensing data can predict pest infestations or disease outbreaks, enabling timely interventions that prevent yield losses. By analyzing soil moisture, temperature, and nutrient levels, satellite technology helps farmers apply fertilizers and water more efficiently. These capabilities reduce waste, increase productivity, and minimize environmental impact. The integration of satellite data into agriculture demonstrates how space technology directly supports global efforts to feed a growing population sustainably.

Satellites also enhance public health and disaster response. Health organizations use satellite imagery to monitor environmental conditions that influence the spread of diseases. Early warning systems for floods, hurricanes, and earthquakes rely on orbital observations to predict impacts and coordinate relief efforts. By providing real-time information about affected areas, satellites enable faster, more effective responses that save lives and reduce damage. Remote sensing data also helps governments plan infrastructure in ways that minimize risk and improve community resilience. The role of satellites in disaster preparedness exemplifies how space technology transforms raw data into actionable solutions for society.

Communication satellites further extend the problem-solving potential of space technology. They enable the transmission of information to remote and underserved regions, supporting education, healthcare, and economic development. Telemedicine relies on these networks to connect specialists with patients in isolated areas, improving access to quality care. Satellite-based internet and communication systems enhance disaster response by providing reliable channels when terrestrial networks fail. Businesses and governments benefit from instantaneous data transfer, facilitating coordination and decision-making on a global scale. These communication capabilities highlight how satellites are more than observational tools—they are integral to solving real-world problems efficiently and equitably.

The technological capabilities of satellites are incredibly diverse, allowing them to address a wide range of global challenges. Optical and radar imaging sensors capture detailed views of Earth's surface, enabling analysis of land use, vegetation, and urban development. Multispectral and hyperspectral sensors detect subtle changes in ecosystems, helping scientists monitor biodiversity and identify early signs of environmental degradation. Satellites equipped with altimeters measure sea level changes, glacier retreat, and river flows, providing critical data for water resource management. By combining multiple sensors and observing platforms, researchers gain comprehensive insights into complex Earth systems. This integrated approach allows for more accurate predictions and informed decision-making at local, national, and global levels.

Agricultural planning and food security benefit significantly from satellite technology. Remote sensing data can identify areas affected by drought, flooding, or nutrient deficiencies, allowing farmers and policymakers to intervene proactively. Satellite-guided precision agriculture reduces overuse of fertilizers and pesticides, conserving resources and protecting ecosystems. Crop yield predictions derived from satellite data help governments plan food distribution and mitigate potential shortages. Additionally, satellites support the tracking of livestock, monitoring grazing patterns, and preventing overuse of pastures. By providing actionable insights, satellites enhance productivity while promoting sustainability in the agricultural sector.

Satellites also play a critical role in environmental conservation and climate change mitigation. Forest monitoring satellites track deforestation, illegal logging, and reforestation efforts, enabling targeted conservation strategies. Ocean-monitoring satellites measure sea surface temperatures, currents, and algal blooms, protecting marine ecosystems and fisheries. Greenhouse gas monitoring satellites provide essential data for climate policy, verifying emissions reduction commitments and guiding international agreements. By continuously observing ecosystems from space, satellites allow scientists and policymakers to detect changes early and implement timely interventions. This proactive approach is essential for maintaining biodiversity and mitigating the long-term impacts of climate change.

Public health initiatives increasingly rely on satellite data to understand and control disease spread. Satellites provide information on environmental conditions that affect vector-borne diseases, such as malaria or dengue fever. By mapping temperature, humidity, and rainfall patterns, health authorities can predict outbreak zones and allocate resources more effectively. Air quality monitoring from satellites also identifies pollution hotspots, informing policies that protect vulnerable populations. Furthermore, disaster response benefits from satellite imagery by assessing infrastructure damage, population displacement, and access routes for aid delivery. Satellites, therefore, support proactive, data-driven strategies that save lives and improve public health outcomes.

Global communication networks enabled by satellites enhance both daily life and emergency response. Satellite internet provides connectivity to remote areas, supporting education, healthcare, and economic opportunities. During natural disasters, when terrestrial networks may fail, satellite communication ensures continued coordination and information flow. Telemedicine, remote learning, and virtual collaboration rely on these systems to overcome geographical barriers. Businesses and governments also use satellite networks for logistics, monitoring, and security, improving operational efficiency. By combining observational and communication capabilities, satellites form a backbone for solving complex societal challenges in real time.

Satellites have become indispensable for predicting and managing natural disasters. By monitoring weather systems, seismic activity, and ocean conditions, satellites provide early warnings for hurricanes, tsunamis, and floods. This data allows governments and emergency agencies to prepare evacuation plans, mobilize resources, and reduce human and economic losses. Remote sensing also helps assess the extent of damage immediately after a disaster, guiding rescue and relief operations. With real-time updates from orbit, response teams can coordinate more efficiently and make informed decisions under pressure. The predictive power of satellites transforms disaster management from reactive to proactive, saving countless lives worldwide.

Water resource management has been revolutionized by satellite observations. Sensors can monitor rainfall, river flows, groundwater levels, and soil moisture across vast regions. This information enables better planning for agriculture, urban water supply, and hydropower generation. Satellite data also identifies regions at risk of drought or flooding, allowing authorities to implement mitigation strategies in advance. By tracking water distribution and consumption, satellites support sustainable management of one of Earth's most critical resources. These insights help prevent scarcity, protect ecosystems, and ensure equitable access to water for communities. Space-based monitoring, therefore, plays a vital role in maintaining global water security.

Environmental monitoring is one of the most significant contributions of satellites to solving Earthly problems. Satellites track deforestation, desertification, ice sheet melting, and ocean health, providing critical data to scientists and policymakers. Air and water quality can be assessed remotely, allowing for targeted interventions to reduce pollution. Greenhouse gas emissions are measured with high precision, informing climate policy and international agreements. By continuously observing the planet, satellites create a comprehensive and dynamic picture of environmental change. This enables decision-makers to act quickly, mitigating damage and promoting sustainability.

Satellites also contribute to urban planning and infrastructure management. High-resolution imagery allows city planners to analyze land use, transportation networks, and population distribution. This data supports the development of efficient public transit, disaster-resilient infrastructure, and sustainable urban growth. Additionally, satellites assist in monitoring energy usage and environmental compliance, promoting greener cities. By providing accurate, up-to-date information, satellites help balance human development with environmental preservation. These capabilities illustrate the multifaceted role of satellites in addressing complex, interconnected challenges on Earth.

Satellites have transformed modern agriculture by providing precise, real-time information that optimizes farming practices. By analyzing soil moisture, temperature, and nutrient levels, farmers can make informed decisions about irrigation, fertilization, and planting schedules. Satellite data helps identify crop stress, pest infestations, or disease outbreaks early, allowing timely interventions that prevent yield loss. Large-scale monitoring enables governments and agricultural organizations to forecast food production, manage supply chains, and ensure food security. Precision agriculture reduces resource waste, lowers environmental impact, and increases productivity. By applying satellite technology to agriculture, humanity can feed a growing population while protecting the environment.

In climate science, satellites provide critical data for understanding and mitigating global warming. They measure greenhouse gas concentrations, track ocean currents, monitor ice sheet dynamics, and observe atmospheric changes over time. This information allows scientists to model future climate scenarios, anticipate extreme weather events, and design strategies for mitigation and adaptation. Governments rely on satellite data to inform climate policies, emission regulations, and international agreements. By providing objective, real-time observations, satellites reduce uncertainty and guide effective environmental action. The ability to monitor climate change continuously from orbit strengthens humanity's capacity to protect ecosystems and human communities alike.

Satellites also play a pivotal role in global health monitoring and disease prevention. Remote sensing data helps track environmental conditions that influence the spread of infectious diseases, such as rainfall patterns, temperature, and vegetation cover. Health organizations use this information to predict outbreak zones and allocate resources efficiently. Satellite-enabled communication supports telemedicine, remote diagnostics, and health education in regions lacking healthcare infrastructure. Additionally, satellites monitor air pollution and other environmental hazards that affect human health. By integrating satellite technology into public health strategies, governments and organizations can prevent illness and respond more effectively to emerging threats.

Economic development and resource management benefit greatly from satellite data. Governments and businesses use orbital observations to monitor natural resources, identify optimal locations for agriculture, energy, or infrastructure, and prevent overexploitation. Satellite imagery supports sustainable forestry, fisheries management, and mineral exploration, balancing economic growth with ecological protection. Remote sensing reduces operational risks and increases efficiency by providing accurate, large-scale information quickly. These capabilities empower nations to make informed decisions that foster long-term prosperity and environmental stewardship. Satellites thus serve as vital tools for promoting both economic resilience and sustainability.

The transformative impact of satellites on Earth is evident across environmental, economic, and social domains. By providing precise, real-time data, satellites enable humanity to understand complex systems and respond effectively to challenges. Their contributions to disaster management, agriculture, climate monitoring, and public health demonstrate the breadth and depth of their influence. Satellites empower governments, organizations, and communities to make informed decisions, optimize resources, and protect vulnerable populations. They bridge gaps in knowledge and connectivity, ensuring that solutions are equitable and globally relevant. The ability to observe, analyze, and communicate from orbit makes satellites indispensable tools for solving some of the world's most pressing problems.

In environmental management, satellites allow continuous monitoring of Earth's ecosystems, helping to preserve biodiversity and mitigate the effects of climate change. Deforestation, desertification, and pollution can be tracked in real time, enabling targeted interventions that reduce damage and promote restoration. Ocean health, water resources, and atmospheric conditions are monitored with unprecedented accuracy, providing insights that guide sustainable development. This information also informs international climate agreements, policies, and mitigation strategies, ensuring coordinated global action. By linking observation to intervention, satellites create a feedback loop that strengthens humanity's ability to manage natural resources responsibly.

Satellites continue to enhance global health and safety initiatives. Remote sensing supports predictive models for disease outbreaks, enabling timely interventions that save lives. Air quality monitoring informs regulations that protect populations from pollution-related illnesses. Disaster response is more efficient, as relief organizations can coordinate logistics, assess damage, and prioritize aid using satellite imagery. These capabilities reduce risks, save resources, and improve resilience in communities worldwide. Satellites, therefore, function as both early warning systems and strategic planning tools for global health and safety.

The economic benefits of satellites are equally profound. By providing accurate data on resources, infrastructure, and market conditions, satellites optimize decision-making and reduce operational risks. Industries such as agriculture, transportation, energy, and insurance rely heavily on satellite-derived information to increase efficiency and profitability. Remote sensing reduces costs associated with errors, delays, and resource mismanagement. Satellite communications open access to education, healthcare, and business opportunities for underserved populations. Through these applications, satellites create a virtuous cycle of innovation, sustainability, and economic development that impacts billions of lives.

Ultimately, satellites exemplify how space technology can be harnessed to solve Earth's greatest challenges. Their observational, analytical, and communication capabilities provide solutions that extend far beyond their initial purpose. By bridging scientific knowledge with practical application, satellites enable humanity to manage natural resources, protect public health, and respond to disasters with unprecedented precision. International collaboration on satellite missions strengthens global cooperation, ensuring that the benefits of space technology are shared equitably. The use of satellites as tools for global problem-solving highlights the enduring potential of space innovation to improve life on Earth. As we continue to advance satellite technology, its role in sustaining a healthier, safer, and more connected planet will only grow.

Chapter . 4

Space-Based Communication for Global Connectivity

Showing how communication satellites can bridge digital divides and bring reliable internet to underserved regions.

Space-based communication has revolutionized the way people connect across the globe. Satellites provide instantaneous voice, video, and data transmission, overcoming geographical barriers that previously limited connectivity. Remote and underserved regions, which lacked traditional infrastructure, now have access to education, healthcare, and economic opportunities. Satellite networks enable real-time communication for governments, businesses, and individuals, fostering collaboration and innovation. By creating a global communication infrastructure, satellites help bridge the digital divide and ensure equitable access to information. The transformative impact of these technologies demonstrates how space-based solutions address social and economic challenges on Earth.

The evolution of communication satellites began with simple relay systems and has advanced to complex constellations providing near-global coverage. Early satellites transmitted basic signals, connecting a few cities and institutions, but they demonstrated the potential for broader applications. Modern satellite networks support high-speed internet, television, telemedicine, and remote learning, reaching millions of people worldwide. Innovations in antenna design, bandwidth optimization, and signal processing have improved reliability, speed, and accessibility. These systems are now critical for emergency communication, military coordination, and disaster response. The continued development of space-based communication networks ensures that no region is left disconnected in an increasingly digital world.

One of the most significant benefits of satellite communication is its ability to support remote education and telemedicine. Students in isolated areas can access online courses, libraries, and virtual classrooms, expanding educational opportunities. Healthcare providers use satellite-enabled telemedicine to consult with specialists, monitor patients remotely, and deliver life-saving treatments in areas without hospitals. These technologies reduce inequality in access to critical services and empower communities to improve their quality of life. By overcoming physical barriers, satellites provide a platform for social and economic development in regions that were previously marginalized. Communication satellites, therefore, serve as tools for both connectivity and empowerment.

Satellite networks also enhance disaster response and emergency management. When natural disasters or conflicts disrupt terrestrial communication systems, satellites provide resilient, reliable channels for coordination. Relief agencies can communicate with field teams, track resources, and direct aid efficiently, saving lives and reducing damage. Early warning systems for hurricanes, floods, and wildfires rely on satellite networks to transmit real-time alerts to communities at risk. These capabilities enable governments and organizations to respond quickly, minimize impact, and maintain critical services. By providing robust communication channels, satellites strengthen global resilience and disaster preparedness.

Economic and commercial benefits of satellite communication are extensive. Businesses rely on satellites for banking, logistics, and international trade, allowing real-time monitoring and secure communication across continents. Internet service providers use satellite networks to expand coverage to rural and remote areas, supporting entrepreneurship and economic growth. Industries such as media, finance, and transportation have been transformed by satellite-enabled data and communication systems. Additionally, the satellite industry itself drives innovation, creating jobs in engineering, manufacturing, and data analysis. By connecting people, resources, and markets, space-based communication serves as a backbone for global development and prosperity.

Satellite internet has become a critical tool for bridging the digital divide across the globe. In remote and rural areas, where laying fiber-optic cables is cost-prohibitive, satellite networks provide reliable connectivity. Students can attend online classes, access educational materials, and participate in virtual learning communities, regardless of location. Healthcare providers use satellite connections to offer telemedicine services, remotely diagnose patients, and provide specialized care that would otherwise be inaccessible. Businesses gain access to global markets, enabling entrepreneurship, e-commerce, and economic development in underserved regions. By ensuring universal connectivity, satellite internet reduces inequalities and creates opportunities for social and economic advancement.

Communication satellites also support emergency and disaster response in critical ways. During earthquakes, hurricanes, floods, and wildfires, terrestrial communication networks are often damaged or destroyed. Satellite systems provide backup connectivity, allowing rescue teams to coordinate operations and communicate with affected populations. Real-time data transmission from drones, sensors, and field teams ensures that relief efforts are efficient and targeted. Early warning alerts can be delivered quickly to communities at risk, reducing casualties and property damage. By offering reliable communication in crisis situations, satellites enhance resilience and save lives in vulnerable regions.

Educational initiatives have been transformed by space-based communication technology. Schools in remote villages can connect with teachers, experts, and peers around the world. Distance learning platforms powered by satellite networks enable students to pursue higher education without relocating to urban centers. Educational content, interactive lessons, and skill-building programs are delivered seamlessly, improving literacy and employment opportunities. Satellite-enabled communications also support teacher training, professional development, and curriculum sharing across borders. This connectivity fosters knowledge exchange, collaboration, and innovation, equipping students with the tools to succeed in a modern, digital economy.

Healthcare delivery benefits immensely from satellite communication networks. Remote clinics and field hospitals can consult specialists in real time, conduct diagnostics, and monitor patient health via telemedicine. Public health organizations use satellite-enabled systems to track disease outbreaks, coordinate vaccination campaigns, and respond to emergencies. Mobile health units rely on satellite connectivity for navigation, data reporting, and communication with central facilities. These technologies ensure that quality healthcare reaches populations in isolated or conflict-affected regions. By connecting healthcare providers and communities, satellites reduce disparities and improve outcomes worldwide.

The economic impact of satellite communication extends to global commerce and market participation. Small businesses and entrepreneurs in rural areas can access banking, e-commerce platforms, and digital payment systems through satellite networks. Logistics and supply chain operations are optimized using real-time communication and tracking enabled by satellites. Industries such as finance, agriculture, media, and transportation benefit from faster, more reliable data transmission and decision-making. By enabling global connectivity, satellite networks foster economic inclusion, innovation, and growth. This integration of space technology into commerce demonstrates how communication satellites drive prosperity alongside social and educational progress.

Satellite communication plays a vital role in disaster management and emergency coordination. When earthquakes, floods, or hurricanes strike, terrestrial communication networks often fail, leaving communities isolated. Satellite networks provide resilient, immediate channels for emergency responders to communicate, coordinate, and deliver aid effectively. Real-time satellite data helps map affected areas, track infrastructure damage, and identify safe routes for relief operations. Early warning alerts transmitted via satellite enable people to evacuate before disasters strike, reducing casualties. By maintaining connectivity under extreme conditions, satellites ensure that emergency services can operate efficiently, saving lives and minimizing property loss.

Governments and international organizations increasingly rely on satellite communication to support remote governance and administrative services. In regions with limited physical infrastructure, satellites facilitate teleconferencing, secure data transmission, and e-government services. Citizens can access essential services, legal information, and public announcements even in hard-to-reach areas. This connectivity strengthens transparency, accountability, and citizen engagement, fostering more effective governance. Satellite communication also supports cross-border cooperation in trade, security, and humanitarian aid. By enabling efficient administration and oversight, satellites strengthen the foundation of governance in both urban and remote regions.

Satellite networks have become essential for global collaboration and international initiatives. Scientific research, environmental monitoring, and climate data sharing rely on communication satellites to transmit information between countries. Space-based networks allow organizations to coordinate efforts in disaster response, conservation, and development projects. Collaborative satellite projects, such as constellations designed for Earth observation, illustrate how shared resources can provide collective benefits. These systems facilitate knowledge exchange and innovation across borders, enhancing global problem-solving capacity. Satellites, therefore, not only connect people but also foster cooperation and collaboration on an international scale.

The infrastructure of satellite communication has advanced rapidly with the deployment of constellations and high-throughput satellites. Modern networks provide low-latency, high-speed internet and coverage in previously unreachable regions. These advancements expand the potential for telemedicine, e-learning, remote business operations, and cultural exchange. Constellations of smaller satellites also allow redundancy, resilience, and continuous coverage in diverse geographic areas. As a result, societies can rely on stable communication networks for both daily operations and crisis response. This technological evolution demonstrates how space-based communication infrastructure can underpin sustainable global connectivity.

Modern communication satellites have undergone significant technological advancements that expand their global impact. High-throughput satellites (HTS) provide increased bandwidth, enabling faster internet speeds and supporting more users simultaneously. Low Earth orbit (LEO) constellations reduce latency and improve coverage in remote and underserved regions, making real-time communication more reliable. Advanced antennas, signal processing, and network management systems enhance the efficiency and resilience of these satellite networks. These technological improvements make it possible to deliver high-quality education, telemedicine, and business services to areas previously left disconnected. The evolution of satellite technology demonstrates that continuous innovation is key to achieving global connectivity.

LEO satellite constellations are particularly transformative for bridging digital divides. By orbiting closer to Earth, these satellites reduce communication delays, enabling applications that require real-time data, such as video conferencing and remote surgery. Constellations also provide redundancy, ensuring uninterrupted service even if one satellite experiences technical issues. With dozens or even hundreds of satellites working together, operators can provide seamless coverage to regions without traditional infrastructure. This approach allows governments, businesses, and humanitarian organizations to extend their reach, improving education, healthcare, and economic opportunities worldwide. LEO constellations illustrate how satellite technology can deliver both speed and accessibility on a global scale.

High-throughput satellites improve capacity and efficiency for densely populated and high-demand areas. These satellites use multiple spot beams and frequency reuse to maximize data throughput. As a result, urban centers, industrial hubs, and rural communities alike can access reliable internet and communication services. This capacity enables tele-education, remote work, e-commerce, and virtual collaboration on a scale never before possible. High-throughput satellites also reduce operational costs by providing more data per satellite, making connectivity more affordable. By combining coverage, speed, and efficiency, modern satellite technology addresses both the accessibility and quality aspects of global communication.

Satellite communication systems also enhance the resilience of global networks. By providing backup connectivity during natural disasters, conflicts, or infrastructure failures, satellites ensure continuity of critical services. Redundant networks allow emergency responders, healthcare providers, and governments to maintain communication when terrestrial systems fail. This reliability is essential for disaster preparedness, humanitarian operations, and maintaining public safety. Additionally, satellite networks support secure communication for military, scientific, and economic applications, protecting sensitive information while promoting collaboration. The resilience of space-based communication networks underscores their importance in maintaining global stability and connectivity.

The transformative impact of space-based communication is evident across social, economic, and humanitarian domains. By bridging digital divides, satellites provide access to education, healthcare, and economic opportunities for populations in remote and underserved regions. Students can participate in virtual classrooms, professionals can consult remotely with experts, and communities gain access to critical information that improves daily life. Satellite communication also supports disaster preparedness and response, ensuring that early warnings, emergency coordination, and relief efforts reach those in need promptly. These capabilities save lives, reduce economic losses, and enhance resilience in vulnerable communities worldwide. Space-based networks, therefore, play a pivotal role in promoting equitable access to information and services.

Economic growth and global commerce are further accelerated by satellite communication networks. Businesses in remote areas can connect to international markets, conduct digital transactions, and engage in e-commerce without traditional infrastructure. Supply chains benefit from real-time tracking and coordination enabled by satellite data, reducing delays and increasing efficiency. Industries such as finance, media, transportation, and energy rely on satellite networks for secure, high-speed communication. By enabling participation in the global economy, satellite technology promotes innovation, entrepreneurship, and prosperity. The economic benefits of connectivity demonstrate that satellites are not only tools for communication but also engines of development.

Satellite communication strengthens governance and international collaboration. Governments can reach citizens in remote regions, deliver essential services, and maintain transparency and accountability. Communication networks support cross-border cooperation for disaster management, climate monitoring, and humanitarian aid. Shared access to satellite data encourages knowledge exchange and coordinated action on global challenges. By enabling collaboration and connectivity, satellites foster diplomatic relationships, international research, and collective problem-solving. Space-based communication, therefore, contributes to both national development and global stability.

The social and cultural impact of satellites is also profound. By connecting communities across vast distances, satellites foster understanding, knowledge sharing, and cultural exchange. People gain access to news, entertainment, and educational content that would otherwise be unavailable. Telecommunication enables remote collaboration, strengthens family ties, and empowers communities to participate in civic and economic life. These networks help reduce inequality, promote inclusivity, and build social cohesion. Satellite communication demonstrates that access to information and connectivity is a fundamental driver of human development and societal well-being.

Ultimately, space-based communication exemplifies the power of satellites to solve real-world problems. By bridging digital divides, supporting education, enabling healthcare, and promoting economic and social development, satellites transform lives and strengthen communities. Their role in disaster response, governance, and international collaboration highlights the broad utility and impact of these technologies. The ongoing evolution of satellite networks ensures that connectivity will continue to reach new populations, foster innovation, and support sustainable development worldwide. Space-based communication, therefore, is a cornerstone of a connected, equitable, and resilient global society, demonstrating how space technology can directly improve life on Earth.

Chapter . 5

International Collaboration in Space as a Model for Earth

Examining how space partnerships can inspire global cooperation to address environmental and humanitarian crises.

International collaboration in space has created a blueprint for solving complex global challenges. Programs such as the International Space Station (ISS) demonstrate how countries with different political, economic, and cultural backgrounds can work together toward a shared goal. By pooling resources, expertise, and technology, these partnerships achieve outcomes that would be impossible for any single nation. Cooperation in space requires mutual trust, shared responsibilities, and long-term planning—principles that can be applied to environmental, humanitarian, and scientific initiatives on Earth. The lessons learned from these partnerships show that collaborative problem-solving is essential for addressing global crises. Space exploration thus offers a model for unity, innovation, and collective action.

Scientific research in orbit benefits from international collaboration, producing knowledge that serves the global community. Experiments conducted on joint missions often address universal challenges, such as climate change, human health, and material sciences. Sharing findings and technologies ensures that all participating nations—and even the wider world—gain from these investments. Collaboration reduces duplication of effort, maximizes efficiency, and accelerates technological progress. It also encourages the standardization of procedures, safety protocols, and scientific methodologies. By demonstrating the tangible benefits of cooperative research, space partnerships inspire similar approaches to solving Earth-bound problems.

International space programs provide lessons in resource sharing and logistical coordination. Launch facilities, mission control centers, and research laboratories are often distributed among multiple countries, requiring meticulous planning and communication. This level of coordination fosters diplomatic dialogue, conflict resolution, and compromise. Participating nations learn to negotiate resource allocation, establish priorities, and jointly solve technical challenges. These practices can be translated to global issues such as disaster relief, food security, and energy management. By showing that nations can work together to achieve ambitious goals, space collaboration builds confidence in global cooperation for humanity's collective benefit.

Space partnerships also demonstrate the importance of joint governance and policy frameworks. Agreements on safety, data sharing, orbital debris management, and intellectual property ensure that international missions operate smoothly and fairly. These frameworks are examples of how shared standards and rules can prevent conflict, optimize efficiency, and protect shared resources. Applying similar governance structures to global environmental, economic, and humanitarian challenges could enhance fairness, transparency, and effectiveness. Space collaboration shows that complex systems require clear agreements, monitoring, and accountability to succeed. This provides a model for managing shared resources and responsibilities on Earth.

The cultural and social dimensions of international space collaboration are equally significant. Astronauts and scientists working together develop mutual respect, understanding, and teamwork that transcends national boundaries. These personal and institutional connections promote empathy, reduce misunderstandings, and encourage peaceful relations. Space missions inspire the public worldwide, demonstrating that cooperation can lead to extraordinary achievements. By highlighting shared human goals rather than national competition, international space projects cultivate a sense of global community. This spirit of collaboration offers a powerful example for tackling shared challenges on Earth, from climate change to humanitarian crises.

International collaboration in space relies on well-structured mechanisms that enable countries to work together efficiently. Joint missions often involve multiple nations contributing technology, funding, and personnel, requiring clear agreements on roles and responsibilities. Shared research data ensures that discoveries benefit all participants and can be applied to global challenges beyond the mission itself. Funding strategies often combine national budgets, private investments, and international grants, distributing costs while maximizing the value of research. Technology exchange allows countries to access innovations they may not have developed independently, accelerating progress for all involved. These mechanisms demonstrate that carefully designed collaboration can achieve ambitious goals more effectively than isolated efforts.

Data sharing is a cornerstone of space partnerships, creating transparency and fostering trust among nations. Satellite imagery, experimental results, and telemetry are often shared openly to maximize scientific value. Open access to data allows researchers to collaborate on climate monitoring, disaster response, and health-related studies. This approach not only improves the quality of research but also builds confidence that all parties benefit equitably. Data-sharing agreements encourage accountability, ensuring that investments yield tangible benefits for participants and the wider global community. The success of these practices illustrates how transparency and cooperation can resolve complex, cross-border challenges on Earth.

Joint funding strategies in international space projects provide lessons for global problem-solving. Shared financial responsibility distributes costs while reducing the burden on individual nations. Collaborative funding encourages long-term commitment and risk-sharing, allowing ambitious projects to move forward even when resources are limited. These approaches demonstrate that pooling resources enables innovation that exceeds the capacity of a single country. Funding partnerships also incentivize accountability and efficient project management. Applying these financial strategies to Earth-focused challenges, such as renewable energy, disaster relief, and infrastructure development, could enhance efficiency and global impact.

Technology exchange between countries strengthens capacity and accelerates innovation. By sharing research, engineering techniques, and best practices, participating nations benefit from each other's expertise. Emerging space programs gain access to advanced technologies, while established programs receive fresh perspectives and collaborative input. This mutual exchange promotes global equality in technological development, preventing monopolization and fostering inclusive growth. Lessons from space technology transfer can be applied to Earth-focused challenges, such as sustainable agriculture, energy efficiency, and disaster mitigation. By embracing collaboration, countries can solve complex problems faster, more efficiently, and more equitably.

Effective communication and coordination are essential for international space collaboration. Teams across different time zones, languages, and cultures must synchronize activities, share information, and resolve conflicts. Advanced communication networks, protocols, and project management tools ensure smooth operations. Lessons in coordination and conflict resolution from space programs can inform global initiatives such as climate agreements, humanitarian aid, and multinational infrastructure projects. By demonstrating that diverse teams can work together effectively toward common objectives, space collaboration provides a model for addressing the interconnected challenges facing humanity. The structured, disciplined approach of space partnerships illustrates that cooperation is both achievable and highly beneficial.

The International Space Station (ISS) is one of the most prominent examples of successful global collaboration in space. Built and maintained by a coalition of nations, the ISS demonstrates how shared goals can unite countries with differing political and economic systems. Each partner contributes modules, technology, and expertise, creating a complex but functional orbiting laboratory. The project requires meticulous coordination, trust, and long-term commitment, providing valuable lessons for cooperative efforts on Earth. Research conducted on the ISS addresses universal challenges, including human health, climate monitoring, and advanced materials. By working together in space, countries demonstrate that collaboration can produce results far beyond the capacity of any single nation.

Joint satellite missions further illustrate the power of international cooperation. Programs like the European Space Agency's Sentinel satellites or joint climate-monitoring missions combine expertise, funding, and technology from multiple countries. These satellites provide comprehensive data for climate research, disaster preparedness, and resource management. Collaborative projects allow countries to share costs and reduce duplication of effort while maximizing scientific and societal benefits. By pooling resources and knowledge, nations can address global challenges such as climate change, deforestation, and food security more effectively. These missions show that shared technological goals create tangible benefits for humanity.

Multinational research projects in space foster innovation through diverse perspectives and expertise. Scientists and engineers from different countries contribute unique approaches, methods, and problem-solving strategies. Collaborative research accelerates the development of new technologies, from robotics to advanced materials and medical devices. Shared laboratories, data analysis centers, and remote experimentation facilities enable real-time collaboration across borders. By leveraging diverse skills and knowledge, international projects produce more comprehensive and effective solutions. These initiatives highlight that diversity in collaboration enhances creativity, efficiency, and impact in solving global challenges.

Space exploration partnerships also promote global education and workforce development. Collaborative programs provide training opportunities for students, engineers, and scientists worldwide. Exchange programs, joint research, and internships foster knowledge transfer and develop a new generation of global problem solvers. Educational outreach from these projects inspires young people to pursue careers in STEM fields, ensuring the continuity of expertise necessary for future innovation. By connecting individuals and institutions across nations, space programs cultivate a global community of skilled professionals dedicated to addressing complex challenges. The educational dimension of collaboration ensures that the benefits of space exploration extend far beyond technological achievements.

International agreements on space policy and governance serve as a model for Earth-focused collaboration. Treaties and agreements regulate orbital resources, data sharing, safety protocols, and intellectual property. These frameworks ensure fairness, accountability, and conflict resolution among participating nations. Similar approaches can be applied to global environmental policy, climate agreements, and resource management initiatives. By establishing clear rules and standards, countries can work together effectively while avoiding disputes and inefficiencies. Space collaboration demonstrates that shared governance structures are essential for managing complex, multinational challenges.

Lessons from international space collaboration can be applied directly to environmental protection on Earth. Shared satellite missions, like those monitoring deforestation or greenhouse gas emissions, illustrate the benefits of combining resources and expertise. Countries can pool funding, technology, and research capacity to address global environmental challenges that no single nation can manage alone. Data-sharing agreements and coordinated observation strategies improve the accuracy and scope of environmental monitoring. By learning from the structure and success of space partnerships, nations can implement joint strategies to combat climate change, conserve biodiversity, and manage natural resources. This approach demonstrates that cooperative solutions are often more effective and equitable than isolated efforts.

Disaster relief operations benefit from strategies modeled on space collaboration. Just as multinational teams coordinate complex space missions, global agencies can collaborate to respond to natural disasters with precision and efficiency. Shared communication networks, satellite data, and logistical planning allow multiple countries to contribute resources and expertise quickly. This coordination reduces duplication of effort, ensures aid reaches the most vulnerable populations, and maximizes the impact of relief operations. Lessons from space programs, such as real-time data sharing and integrated command structures, can improve the speed and effectiveness of humanitarian responses. Applying these principles ensures that disaster relief is both collaborative and optimized for impact.

International space cooperation also informs global public health initiatives. The collaborative framework used in space missions, including shared research, standardized protocols, and data transparency, can guide multinational health programs. Satellite data, developed and shared internationally, helps track disease outbreaks, monitor air quality, and manage health resources. Countries can work together to combat pandemics, coordinate vaccination campaigns, and implement preventive measures efficiently. By modeling global health initiatives on successful space collaborations, nations can improve outcomes while reducing inequalities. The principle of shared responsibility and mutual benefit ensures that public health solutions reach everyone, not just a select few.

Climate adaptation strategies benefit from the lessons of space partnerships. Collaborative observation, data analysis, and policy development in space provide a framework for joint climate action on Earth. Countries can share climate projections, environmental models, and adaptive strategies to manage risks like floods, droughts, and sea-level rise. Joint projects allow for coordinated implementation of infrastructure improvements, disaster preparedness programs, and resource management policies. The trust, transparency, and accountability cultivated in space collaborations are critical for ensuring that climate initiatives are equitable and effective. This cooperative approach strengthens resilience across regions, helping communities prepare for and respond to environmental challenges.

International collaboration in space provides a powerful model for solving global challenges on Earth. By demonstrating that nations with diverse interests and resources can work together toward shared goals, space partnerships offer lessons in diplomacy, trust, and joint problem-solving. Coordinated missions, shared data, and pooled resources highlight the value of collaboration over competition in achieving outcomes that benefit humanity as a whole. The successes of the ISS, joint satellite programs, and multinational research projects show that complex challenges can be addressed efficiently when nations unite their expertise and technology. These lessons can inspire cooperative approaches to environmental protection, disaster relief, public health, and sustainable development worldwide. By fostering collaboration, space initiatives teach that collective effort is often more effective than isolated action.

The principles of international space collaboration—trust, transparency, accountability, and shared responsibility—are applicable across many sectors on Earth. Collaborative governance structures used in space missions can inform global agreements on climate change, resource management, and humanitarian aid. The mechanisms for data sharing, joint research, and technology transfer in space provide templates for cooperation in science, health, and environmental policy. By replicating these structures, nations can address transnational issues more effectively, ensuring that solutions are equitable and sustainable. Space collaboration demonstrates that clear agreements, shared goals, and structured coordination are essential for achieving ambitious objectives. These frameworks empower humanity to tackle challenges that no single country can solve alone.

The educational and workforce benefits of space collaboration extend to Earth-focused problem-solving. Multinational programs cultivate a new generation of scientists, engineers, and policymakers who are skilled in teamwork, negotiation, and cross-cultural cooperation. These professionals bring the lessons learned from space projects into other global initiatives, applying collaborative thinking to issues like climate adaptation, food security, and disaster preparedness. By fostering skills in collaboration and innovation, international space programs ensure that future leaders are prepared to solve complex challenges collectively. This transfer of knowledge demonstrates the broader societal value of space partnerships beyond technology and research.

Space collaboration also reinforces the importance of long-term vision and commitment. Joint missions often span decades, requiring sustained coordination, funding, and strategic planning. These long-term efforts demonstrate that global challenges on Earth, such as climate change and humanitarian crises, also require patience, persistence, and cooperative planning. By observing the success of collaborative space projects, nations can approach Earth-focused problems with a similar mindset, emphasizing continuity, shared responsibility, and resilience. This long-term perspective ensures that solutions are not only immediate but also sustainable for future generations.

Ultimately, international collaboration in space illustrates that humanity can achieve extraordinary results when working together. Shared goals, coordinated strategies, and mutual trust produce innovations, discoveries, and practical solutions that benefit all of society. The lessons learned from space partnerships can guide Earth-focused initiatives, from environmental protection and disaster relief to public health and sustainable development. By embracing the principles of collaboration demonstrated in space, nations can address global challenges more effectively, equitably, and sustainably. Space exploration thus serves as both a literal and metaphorical model for the collective problem-solving needed to secure a better future for our planet and all its inhabitants.

Part II: Earth Observation and Environmental Solutions

Chapter . 6

Monitoring Climate Change from Orbit

Explaining how satellite observation enables precise tracking of climate patterns to guide mitigation strategies.

Satellites have transformed the study and management of climate change by providing comprehensive, continuous, and accurate observations of Earth's systems. From measuring atmospheric temperatures and greenhouse gas concentrations to tracking ice sheets and ocean currents, satellites offer data that was previously impossible to obtain at a global scale. These observations allow scientists to detect subtle changes in climate patterns, providing early warnings for extreme weather events. Governments, researchers, and organizations rely on satellite data to inform mitigation strategies, environmental policies, and adaptation plans. By providing a global perspective, satellites ensure that climate actions are based on evidence and not just local observations. This capacity to monitor and analyze Earth from orbit is crucial for managing the planet's changing climate.

Earth observation satellites measure key indicators of climate change, including rising temperatures, sea-level fluctuations, and glacial melting. Instruments onboard satellites detect changes in surface temperature, ice cover, and vegetation, creating a detailed picture of Earth's environmental health. These measurements allow researchers to monitor trends over time, distinguishing between natural variability and human-induced changes. Satellite data feeds into climate models that predict future scenarios, informing policy decisions at national and international levels. By quantifying climate impacts with high accuracy, satellites reduce uncertainty and enable targeted interventions. This precision helps prioritize resources, assess risks, and design effective mitigation strategies.

Satellites also monitor atmospheric composition, including carbon dioxide, methane, and other greenhouse gases. By tracking emissions from industrial, agricultural, and natural sources, scientists can identify trends, hotspots, and areas requiring intervention. These data are essential for evaluating the effectiveness of climate policies, international agreements, and emissions reduction strategies. Satellites also provide early warning of extreme weather events, such as heatwaves, storms, and floods, helping communities prepare and respond. By offering continuous, global coverage, satellites ensure that climate monitoring is consistent and reliable across all regions. This comprehensive view strengthens scientific understanding and informs practical climate solutions.

Ocean observation from space is critical for understanding climate dynamics and their impact on human societies. Satellites measure sea surface temperatures, ocean currents, and wave patterns, which influence weather systems, fisheries, and coastal ecosystems. They also track sea-level rise, ocean acidification, and the melting of polar ice, providing essential information for climate adaptation planning. Coastal communities rely on these insights to develop flood defenses, manage fisheries, and protect ecosystems. By integrating satellite ocean data with atmospheric and terrestrial observations, scientists gain a holistic understanding of Earth's climate system. This interconnected knowledge allows for more accurate predictions and effective climate strategies.

Long-term satellite monitoring provides a continuous record of climate change, allowing researchers to assess trends and project future conditions. Historical satellite data help identify patterns such as accelerating ice melt, shifting rainfall, and rising global temperatures. These datasets are critical for understanding the pace of climate change, its regional variations, and potential impacts on ecosystems and human populations. Policymakers use this information to develop evidence-based adaptation and mitigation strategies that protect vulnerable communities. International organizations also rely on satellite data to coordinate global climate initiatives, track progress, and verify compliance with agreements. By providing an ongoing, comprehensive perspective, satellites are indispensable tools for managing climate change effectively.

Satellite technologies use advanced sensors to capture precise information about Earth's climate. Instruments such as radiometers, spectrometers, and lidar detect changes in temperature, atmospheric composition, and surface reflectivity. These sensors provide high-resolution data on land, ocean, and atmospheric conditions, enabling scientists to study climate trends with unprecedented accuracy. Remote sensing methods allow continuous monitoring over large areas, including regions that are difficult or impossible to access on the ground. By combining data from multiple satellites, researchers can build comprehensive, multi-dimensional models of Earth's climate system. These technological capabilities make satellites indispensable for tracking and understanding the effects of climate change.

Satellites employ both passive and active remote sensing techniques to gather climate data. Passive sensors detect natural radiation emitted or reflected by Earth, such as sunlight reflected from clouds or infrared radiation from the surface. Active sensors, like radar and lidar, emit signals and measure the return after bouncing off Earth's surface, providing detailed information on topography, ice thickness, and vegetation structure. By integrating these methods, satellites capture complementary datasets that reveal both large-scale patterns and localized changes. This combination of approaches enhances the precision and reliability of climate monitoring. Scientists rely on this wealth of information to improve climate models, predict future scenarios, and design mitigation strategies.

Climate satellites also track the concentration and movement of greenhouse gases. Instruments such as spectrometers measure carbon dioxide, methane, and ozone levels in the atmosphere, providing insights into sources, sinks, and global distribution. This data helps policymakers evaluate the effectiveness of emission reduction programs and environmental policies. By identifying regions of high emissions, satellites guide targeted interventions to reduce greenhouse gases. Satellite observations also enable verification of international climate agreements, promoting accountability and cooperation between nations. Accurate monitoring of greenhouse gases is critical for developing strategies that limit global warming and mitigate its impacts.

Satellites monitor changes in Earth's cryosphere, including glaciers, ice sheets, and sea ice. Radar and laser altimetry measure ice thickness and volume, while optical imaging tracks movement and melting patterns. Observing polar regions is essential, as they are highly sensitive indicators of global climate change. Melting ice contributes to sea-level rise, which threatens coastal communities worldwide. Satellites provide early warnings of accelerated ice loss and enable governments to develop adaptation plans. This information is crucial for understanding global climate dynamics and implementing effective mitigation strategies.

Vegetation and land surface monitoring from space contributes to climate science and sustainability. Satellite imagery captures deforestation, desertification, urban expansion, and seasonal vegetation changes. These data help assess carbon sequestration, land use impacts, and ecosystem health. By tracking changes in forests, grasslands, and agricultural areas, satellites inform conservation and resource management strategies. Remote sensing also enables early detection of drought, soil degradation, and other environmental stressors that exacerbate climate change. Integrating land surface data with atmospheric and oceanic measurements provides a comprehensive understanding of Earth's climate system, supporting informed decision-making at all levels.

Satellite climate data is crucial for predicting extreme weather events and protecting communities worldwide. By continuously monitoring atmospheric pressure, temperature, humidity, and wind patterns, satellites allow meteorologists to forecast hurricanes, typhoons, floods, and heatwaves with increasing accuracy. Early warning systems rely on this information to issue timely alerts, helping authorities coordinate evacuations, deploy emergency services, and minimize damage. Satellites also track storm intensity and movement in real time, enabling dynamic responses to evolving situations. Accurate predictions save lives, reduce property loss, and allow governments to allocate resources more effectively. By providing global coverage, satellites ensure that even remote or underdeveloped regions benefit from advanced weather forecasting capabilities.

Climate models rely heavily on data collected from satellites to simulate future scenarios. Observations of greenhouse gas concentrations, ocean temperatures, and ice sheet dynamics feed into sophisticated computational models. These models project potential outcomes such as sea-level rise, shifting precipitation patterns, and extreme weather frequency. Policymakers use these projections to develop mitigation strategies, plan infrastructure, and allocate resources for climate adaptation. Continuous satellite observations refine model accuracy by updating input data and validating predictions. This iterative process ensures that global strategies are informed by the most precise and current scientific information available.

Satellites also provide essential data for international climate agreements and policy-making. By tracking carbon emissions, deforestation rates, and energy usage patterns, they help countries monitor compliance with treaties such as the Paris Agreement. Transparent satellite data builds trust among nations, reducing disputes over environmental responsibility. It also enables verification of progress toward emission reduction goals and sustainable development targets. International organizations rely on these observations to coordinate global efforts and set realistic, measurable objectives. In this way, satellites serve as both a scientific and diplomatic tool in the fight against climate change.

In addition to global monitoring, satellites aid local and regional climate adaptation strategies. High-resolution imagery identifies areas vulnerable to sea-level rise, flooding, drought, and heat stress. This information helps city planners, farmers, and emergency managers develop targeted interventions, such as flood defenses, irrigation planning, and early warning systems. By tailoring climate strategies to local conditions, satellite data ensures more effective risk management and resource allocation. It also informs infrastructure design, land use planning, and conservation efforts. Through these applications, satellites bridge the gap between global observations and practical, on-the-ground solutions.

Satellites play a critical role in monitoring polar ice and assessing the effects of global warming on glaciers and ice sheets. Radar and laser altimetry measure ice thickness, movement, and melting rates with high precision. Observing these regions is essential because polar ice responds quickly to climate change and serves as a key indicator of global temperature trends. Shrinking ice sheets contribute to rising sea levels, threatening coastal populations and ecosystems worldwide. By providing continuous, long-term data, satellites enable scientists to track these changes accurately and predict future impacts. This information is invaluable for guiding climate adaptation and mitigation strategies in vulnerable regions.

Monitoring sea-level rise is another vital application of satellite technology. Satellite altimeters measure ocean height across the globe, detecting subtle increases caused by melting ice and thermal expansion. These measurements allow scientists to forecast coastal flooding, plan urban defenses, and evaluate the risk to infrastructure and populations. Combining sea-level data with climate models helps policymakers prioritize adaptation measures, from building seawalls to relocating communities at risk. Satellite observations also track tides, storm surges, and changes in ocean circulation, which influence local and regional climate patterns. By providing a global perspective on sea-level changes, satellites inform sustainable coastal management and disaster preparedness.

Satellites also measure ocean temperatures and currents, which are critical for understanding climate systems. Sea surface temperature affects weather patterns, marine ecosystems, and the frequency of extreme events such as hurricanes and heatwaves. Satellite data helps predict El Niño and La Niña phenomena, which have global consequences for agriculture, fisheries, and water management. Monitoring ocean currents provides insight into energy transport across the planet, which impacts climate models and long-term forecasting. By tracking these dynamic processes, satellites improve our understanding of how climate change affects both local communities and global systems. Accurate ocean monitoring is essential for informed mitigation and adaptation strategies.

Deforestation and land use changes are tracked effectively with satellite observation. Optical and multispectral imaging allows scientists to measure forest loss, reforestation progress, and habitat changes in near real time. These data help governments, NGOs, and international organizations implement policies to reduce emissions, preserve biodiversity, and promote sustainable land management. Tracking deforestation also supports verification of international climate agreements and encourages accountability. Satellites provide actionable insights that enable targeted interventions, such as enforcing protected areas or guiding sustainable agriculture practices. By linking deforestation data to climate models, satellites ensure that land management strategies contribute effectively to mitigation efforts.

Satellite monitoring of climate change has transformed the world's ability to respond to environmental challenges. By providing comprehensive, continuous, and accurate data, satellites allow scientists, governments, and organizations to understand the pace, scope, and impacts of global warming. This knowledge informs mitigation strategies, including emission reduction programs, renewable energy planning, and sustainable land management. Satellites also guide adaptation measures, from designing resilient infrastructure to preparing communities for extreme weather events. By offering real-time observations, predictive modeling, and long-term datasets, satellites ensure that climate policies are evidence-based and effective. Their role is essential for protecting ecosystems, economies, and human populations worldwide.

International collaboration in satellite climate monitoring enhances global climate action. Shared satellite missions, data exchange, and coordinated observation networks ensure that all countries, regardless of resources, have access to critical climate information. This transparency fosters trust among nations, supports climate agreements, and encourages collective action. Multinational projects combine expertise, funding, and technology to achieve more comprehensive and accurate monitoring than any single country could manage alone. By demonstrating the benefits of cooperation, satellite programs provide a model for addressing global environmental challenges collectively. Collaboration ensures that mitigation and adaptation strategies are informed, equitable, and effective.

Satellite data also drives innovation in climate science and technology. Advanced sensors, remote sensing techniques, and integrated observation systems enable researchers to develop more precise climate models and predictive tools. These innovations improve disaster forecasting, agricultural planning, and ecosystem management. Satellite-derived insights also inform emerging technologies, such as carbon capture, renewable energy optimization, and climate-resilient infrastructure. By linking scientific discovery to practical applications, satellites help translate knowledge into tangible solutions for climate mitigation and adaptation. This integration of technology and observation exemplifies how space systems contribute directly to solving Earth's most pressing problems.

The long-term perspective provided by satellite monitoring is essential for sustainable planning. Continuous observation over decades allows scientists to detect trends, identify anomalies, and refine models for future climate projections. Decision-makers can plan infrastructure, urban development, and environmental policies with a clear understanding of potential risks and opportunities. This long-term data supports proactive rather than reactive responses, reducing economic losses and human suffering. It also ensures that mitigation and adaptation strategies remain effective under changing conditions. Satellites, therefore, provide the continuity and reliability necessary for long-term climate resilience.

Ultimately, monitoring climate change from orbit exemplifies the critical role of space technology in protecting life on Earth. Satellites enable evidence-based decision-making, early warnings, and global cooperation, ensuring that humanity can respond to environmental challenges effectively. By integrating observations of atmosphere, oceans, ice, and land, satellites offer a comprehensive view of climate systems and their interconnections. Their contributions inform mitigation, adaptation, policy, and technological innovation, guiding humanity toward a more sustainable and resilient future. Space-based climate monitoring demonstrates that Earth-focused applications of space technology are essential for safeguarding the planet and its inhabitants for generations to come.

Chapter . 7

Predicting Natural Disasters Before They Strike

Using space technology to forecast earthquakes, hurricanes, and floods, saving lives and reducing damage.

Satellite technology has revolutionized disaster prediction by providing timely, accurate, and comprehensive data on natural hazards. From orbit, satellites monitor atmospheric conditions, seismic activity, and ocean patterns, enabling early detection of potential disasters. This data allows scientists and emergency managers to issue warnings, coordinate evacuations, and plan interventions before events occur. By providing a global perspective, satellites ensure that even remote and underserved areas benefit from predictive capabilities. Governments and organizations rely on satellite observations to prepare communities, allocate resources, and reduce the human and economic toll of disasters. Space-based disaster monitoring demonstrates how technological innovation can directly protect human lives and property.

Hurricane forecasting has improved significantly through satellite observations. Satellites track storm formation, intensity, movement, and precipitation in real time. Data on wind speed, pressure, and sea surface temperature allow meteorologists to predict the trajectory and potential impact of hurricanes with greater accuracy. Early warning systems based on satellite data enable timely evacuations, emergency preparedness, and resource allocation. Coastal communities, in particular, benefit from these predictive capabilities, reducing fatalities and property damage. By combining satellite observations with computational models, forecasters can provide detailed, actionable information to decision-makers and the public.

Flood prediction has also been enhanced by satellite technology. Satellites monitor rainfall, river levels, snowmelt, and soil moisture, providing critical data for anticipating floods. Remote sensing allows authorities to identify vulnerable areas and implement preventive measures such as levees, drainage systems, and evacuation plans. Early detection of flash floods and rising river levels helps communities respond quickly, reducing damage and saving lives. In addition, satellite data informs disaster relief operations by mapping affected regions and guiding resource deployment. By integrating hydrological models with satellite observations, governments can improve the efficiency and effectiveness of flood management strategies. This capability is particularly valuable in regions prone to seasonal flooding and extreme weather events.

Earthquake monitoring benefits from satellites through geodetic and remote sensing technologies. While earthquakes are difficult to predict with absolute certainty, satellites detect subtle ground deformations and stress accumulation along fault lines. Interferometric Synthetic Aperture Radar (InSAR) and GPS measurements provide high-resolution data on tectonic movement and strain, which help identify high-risk areas. This information allows scientists to assess seismic hazards, develop early warning systems, and guide infrastructure planning in earthquake-prone regions. Governments can implement building codes, emergency preparedness drills, and risk reduction measures based on satellite-derived insights. By enhancing understanding of seismic processes, satellites contribute to minimizing casualties and economic losses from earthquakes.

Wildfire prediction and monitoring have also been transformed by satellite technology. Satellites track temperature, vegetation dryness, and atmospheric conditions to identify areas at high risk for fire outbreaks. Infrared sensors detect hotspots and active fires, providing near real-time information for firefighting teams. Satellite imagery helps coordinate response efforts, allocate resources efficiently, and protect communities and ecosystems. Long-term monitoring also informs land management, fire prevention strategies, and climate-related risk assessments. By combining predictive analytics with real-time observation, satellites help prevent wildfires from escalating into catastrophic events.

Integrating satellite data with predictive models enhances disaster forecasting accuracy and timeliness. Satellite observations provide critical inputs for numerical models that simulate weather, seismic activity, and hydrological systems. These models combine historical data, real-time satellite readings, and environmental variables to anticipate potential hazards. Early warning systems use this information to alert authorities and communities, allowing them to take preventive measures before disasters strike. By continuously updating models with satellite data, predictions become more reliable, reducing false alarms and increasing trust in emergency alerts. The integration of space-based observations and computational modeling creates a powerful tool for disaster risk reduction worldwide.

Global coordination is essential to maximize the benefits of satellite-based disaster prediction. International networks, such as the World Meteorological Organization and UN disaster management programs, share satellite data across borders. This collaboration ensures that countries lacking their own space infrastructure can still access life-saving information. Shared data improves regional disaster preparedness, as neighboring countries can anticipate cross-border hazards such as hurricanes, floods, or volcanic eruptions. By pooling resources and expertise, nations can develop standardized protocols, improve forecasting models, and coordinate humanitarian responses. Space technology, therefore, not only predicts disasters but also fosters global cooperation in disaster management.

Satellite data also enhances real-time situational awareness during disaster events. High-resolution imagery and remote sensing provide a dynamic view of affected areas, enabling emergency teams to assess damage quickly. Authorities can track floodwaters, wildfire spread, and landslide activity as they happen, guiding evacuation routes and resource deployment. This capability minimizes response time and optimizes the use of limited emergency resources. By providing a bird's-eye perspective, satellites enable decision-makers to prioritize interventions and allocate aid efficiently. The combination of prediction and real-time monitoring strengthens resilience and saves lives during natural disasters.

In addition to immediate disaster response, satellites contribute to long-term risk assessment and mitigation planning. Data on historical storm tracks, seismic activity, and wildfire frequency helps identify high-risk regions. Urban planners, engineers, and governments use this information to design resilient infrastructure, enforce building codes, and implement zoning regulations. Satellite-based mapping supports hazard assessment for critical infrastructure such as hospitals, schools, and transportation networks. By understanding vulnerability patterns, communities can develop preparedness strategies and reduce long-term risks. Space-based data thus informs proactive measures that minimize the impact of future disasters.

Satellite technology also supports community-level awareness and preparedness programs. Real-time alerts, weather forecasts, and hazard maps derived from satellite data can be shared with local populations. Mobile applications and broadcast systems distribute warnings and safety instructions, empowering individuals to take preventive actions. Community training programs leverage satellite data to simulate disaster scenarios and improve emergency response skills. By combining technology with education, societies can enhance resilience and reduce casualties during extreme events. Satellites, therefore, play a dual role in both forecasting disasters and fostering informed, prepared communities.

Satellite technology improves earthquake preparedness by detecting subtle ground movements and tectonic stress. Interferometric Synthetic Aperture Radar (InSAR) measures minute shifts in the Earth's crust, identifying areas at increased risk of seismic activity. Combined with GPS data, satellites help scientists model fault line behavior and anticipate potential earthquake zones. This information guides urban planning, infrastructure reinforcement, and emergency response strategies. Early identification of high-risk areas enables governments to prioritize public safety measures and allocate resources efficiently. While earthquakes cannot yet be predicted with absolute certainty, satellite monitoring reduces uncertainty and strengthens preparedness efforts.

Hurricane prediction has advanced through satellite monitoring of atmospheric and oceanic conditions. Satellites measure sea surface temperatures, cloud patterns, wind speeds, and barometric pressure, providing crucial data for forecasting storm formation and intensity. Real-time tracking allows meteorologists to predict storm paths, estimate landfall locations, and issue early warnings. These predictions enable timely evacuations, deployment of emergency services, and protection of property and infrastructure. Satellite-based hurricane models are continuously refined with updated observations, improving accuracy over time. By combining observation and modeling, satellites save lives and minimize economic losses during tropical cyclones.

Flood prediction benefits greatly from satellite observation of rainfall, river levels, snowpack, and soil moisture. Remote sensing identifies regions at risk of flash floods, seasonal river flooding, or coastal inundation. Satellite-derived hydrological models allow authorities to simulate potential flood scenarios and implement preventive measures. Early warnings based on satellite data guide evacuations, sandbagging, and deployment of rescue resources. Additionally, satellites provide post-event damage assessments to aid in relief planning and recovery. By monitoring both the onset and aftermath of floods, satellites enhance preparedness and improve resilience for affected communities.

Wildfire prediction and monitoring are enhanced by satellites capable of detecting heat, vegetation dryness, and atmospheric conditions. Infrared sensors identify hotspots before fires become unmanageable, while multispectral imagery assesses fuel levels and environmental vulnerability. Fire management teams use these data to allocate personnel and equipment strategically. Satellites also track fire progression, smoke dispersion, and potential threats to populated areas. Long-term monitoring helps identify trends, inform controlled burns, and guide land management policies. By combining predictive and real-time monitoring capabilities, satellites mitigate wildfire impacts and protect lives, property, and ecosystems.

Volcanic activity prediction benefits from satellite observation of thermal, gas, and surface deformation data. Satellites detect changes in ground elevation, heat flow, and gas emissions that precede eruptions. This information allows authorities to assess eruption risk, plan evacuations, and implement hazard mitigation measures. Continuous monitoring of volcanoes in remote or inaccessible regions ensures that warning systems remain active and reliable. Satellite-based observation complements ground-based seismic monitoring, providing a more complete picture of volcanic activity. By integrating satellite data into disaster planning, governments can reduce casualties, protect infrastructure, and preserve natural environments near active volcanoes.

Satellite networks play a central role in providing real-time monitoring for disaster prediction. Constellations of weather, Earth observation, and communication satellites continuously gather data from around the globe. This constant flow of information allows for immediate analysis of potential hazards, including storms, floods, wildfires, and volcanic activity. By transmitting data in real time to national and international disaster management centers, satellites enable rapid decision-making and resource allocation. The integration of multiple satellite systems ensures redundancy, reliability, and comprehensive coverage across land, sea, and atmosphere. Real-time monitoring transforms disaster response from reactive to proactive, saving lives and reducing economic losses.

Early warning systems are strengthened by satellite observations, providing communities with critical time to prepare. Data on storm intensity, river levels, and seismic activity feed predictive models that trigger alerts to governments and populations. Warning messages can be disseminated through mobile networks, broadcast channels, and online platforms, reaching millions within minutes. These systems guide evacuations, emergency supply deployment, and public safety measures. Satellite-enabled early warnings also help protect critical infrastructure, including power grids, hospitals, and transportation networks. By providing timely, actionable information, satellites reduce vulnerability and enhance community resilience to natural hazards.

International cooperation amplifies the effectiveness of satellite-based disaster prediction. Shared satellite data and observation resources allow countries to anticipate hazards that may cross borders. Regional coordination ensures that early warnings are consistent, accurate, and timely for all affected nations. Collaborative networks also facilitate joint research, model development, and training programs, enhancing global preparedness. Countries with limited satellite capabilities gain access to critical data, improving equity in disaster management. By promoting shared responsibility and mutual support, international cooperation strengthens global resilience against natural disasters.

Satellite observation supports comprehensive disaster management beyond prediction alone. Post-event assessments use imagery and sensor data to map affected areas, evaluate infrastructure damage, and guide recovery operations. This information helps prioritize relief efforts, allocate resources efficiently, and plan long-term reconstruction. By comparing pre- and post-event satellite data, authorities can refine predictive models and improve future preparedness. Satellite-based monitoring ensures that both immediate response and long-term planning are informed by accurate, reliable information. This holistic approach enhances the effectiveness of disaster management at local, national, and international levels.

Advanced analytics and artificial intelligence (AI) further enhance satellite-based disaster prediction. Machine learning algorithms analyze vast quantities of satellite data to detect patterns, anomalies, and early warning signals that humans might miss. AI models improve the speed and accuracy of predictions, offering real-time guidance for emergency managers and policymakers. Combining satellite observations with AI enhances the precision of hazard forecasts and reduces false alarms. These tools enable proactive, evidence-based responses, improving outcomes for affected populations. By leveraging technology, satellites extend humanity's ability to anticipate and manage natural disasters more effectively than ever before.

Satellite-based disaster prediction provides significant societal, economic, and humanitarian benefits. By offering early warnings for hurricanes, floods, earthquakes, wildfires, and volcanic eruptions, satellites save countless lives and reduce the human toll of natural hazards. Communities can evacuate safely, governments can deploy emergency resources efficiently, and critical infrastructure can be protected. These capabilities also minimize economic losses by allowing businesses, farmers, and industries to prepare for disruptions. Beyond immediate impacts, satellite monitoring informs long-term disaster preparedness, risk reduction, and urban planning strategies. By integrating observation, prediction, and response, satellites transform how societies manage natural hazards.

The global reach of satellite networks ensures that even remote or underdeveloped regions can benefit from disaster prediction. Countries lacking local monitoring infrastructure gain access to critical information through international data-sharing programs. This equitable access empowers vulnerable populations, reduces disparities in disaster resilience, and strengthens global humanitarian efforts. Regional and global coordination, supported by satellite data, allows countries to respond collectively to transboundary hazards such as cyclones, floods, and volcanic eruptions. International collaboration also enhances trust, cooperation, and resource sharing among nations. Satellites, therefore, not only save lives but also foster global solidarity and resilience.

Satellite-enabled disaster prediction reduces uncertainty and improves planning at all levels of society. Governments, emergency agencies, and community leaders can make informed decisions based on reliable, real-time information. Businesses and agricultural sectors can adjust operations to minimize losses and maintain continuity. Insurance companies benefit from improved risk assessment and loss mitigation data. Scientific research relies on satellite observations to refine hazard models and develop innovative prediction technologies. By providing accurate, actionable information, satellites create a foundation for evidence-based disaster management.

Economic resilience is enhanced through the use of satellite data in disaster prediction. Early warnings allow industries to safeguard assets, secure supply chains, and reduce downtime during extreme events. Governments can allocate emergency funding more efficiently and minimize the economic impact on communities. Long-term planning informed by satellite data supports resilient infrastructure, sustainable land use, and disaster-adaptive policies. These benefits extend to global trade, investment, and development by reducing vulnerability to natural hazards. Satellites therefore contribute directly to the economic stability and sustainability of societies around the world.

Ultimately, predicting natural disasters from orbit demonstrates the indispensable role of space technology in protecting life and property on Earth. By combining real-time observation, predictive modeling, early warning systems, and international cooperation, satellites provide a comprehensive disaster management solution. These capabilities save lives, reduce economic losses, and strengthen global resilience to natural hazards. Satellite technology exemplifies how space-based solutions can address some of the most urgent and complex challenges facing humanity. Through continued innovation, collaboration, and investment, satellites will remain central to safeguarding communities, economies, and ecosystems against the growing threat of natural disasters.

Chapter . 8

Mapping Deforestation and Reforestation Progress

Leveraging remote sensing to manage forests and preserve biodiversity.

Satellite technology has transformed the way we monitor forests, enabling detailed mapping of deforestation and reforestation activities across the globe. Remote sensing instruments detect changes in vegetation cover, forest density, and canopy structure, providing near-real-time data to scientists and policymakers. This information is critical for assessing the health of ecosystems, tracking biodiversity, and enforcing conservation policies. By observing large, remote, or inaccessible areas, satellites overcome limitations of ground-based monitoring methods. Governments, NGOs, and international organizations rely on these observations to implement sustainable forest management practices. Satellite monitoring thus plays a crucial role in preserving Earth's forests and mitigating climate change.

Deforestation mapping helps identify illegal logging, land conversion, and forest degradation. High-resolution optical and multispectral imagery allows authorities to detect changes in forest cover, track the rate of tree loss, and pinpoint areas requiring intervention. Satellite data enables timely enforcement of environmental regulations, protecting both biodiversity and the livelihoods of communities that depend on forests. By monitoring deforestation trends over time, governments can evaluate the effectiveness of policies and develop adaptive management strategies. This monitoring also supports global climate initiatives, as forests act as significant carbon sinks. Through satellite observation, deforestation can be addressed proactively rather than reactively.

Reforestation efforts are similarly enhanced by satellite technology. Satellites monitor newly planted areas, track tree growth, and assess vegetation health over time. These observations ensure that restoration projects are effective and that resources are being used efficiently. Remote sensing also helps identify degraded lands suitable for reforestation, guiding strategic planning and investment. International organizations and governments can use satellite data to report progress toward reforestation goals, such as those established under climate agreements. By providing an objective, verifiable record of forest recovery, satellites increase accountability and encourage continued investment in ecological restoration.

Biodiversity conservation is closely tied to forest monitoring through satellites. Forest loss can fragment habitats and endanger countless plant and animal species. Satellite data helps identify critical habitats, monitor species distribution, and assess ecosystem connectivity. Conservationists use this information to prioritize protected areas, design wildlife corridors, and mitigate the impacts of human activity. In addition, satellites can detect environmental stressors such as drought, fires, and disease outbreaks that threaten forest ecosystems. By integrating forest monitoring with biodiversity data, satellite technology supports a holistic approach to ecosystem preservation and climate mitigation.

Long-term satellite observation provides valuable insights into global forest trends. By analyzing historical data, scientists can identify patterns in deforestation, reforestation, and forest degradation. This long-term perspective informs climate models, carbon accounting, and sustainable land-use planning. Policymakers can develop strategies to reduce deforestation, enhance carbon sequestration, and balance economic development with environmental protection. Satellite monitoring ensures transparency and facilitates international cooperation in forest management. Ultimately, leveraging remote sensing for forest mapping demonstrates the vital role of space technology in preserving Earth's ecosystems and combating climate change.

Satellite technology employs a variety of sensors to monitor forests accurately and efficiently. Optical sensors capture visible light reflected from vegetation, enabling detection of forest cover, canopy density, and land-use changes. Multispectral sensors go further by analyzing multiple wavelengths of light, including infrared, to assess vegetation health, moisture content, and growth patterns. Radar sensors penetrate clouds and vegetation layers, providing reliable measurements even in regions with frequent cloud cover or dense canopy. By combining data from multiple sensors, scientists can generate a comprehensive and accurate map of forest conditions. This integration of technologies allows for continuous, large-scale monitoring of both deforestation and reforestation activities.

Optical satellite imagery is particularly useful for identifying areas where forests are being cleared. High-resolution images can detect roads, logging camps, and patches of tree loss, allowing authorities to pinpoint illegal or unsustainable practices. Time-series analysis of optical images helps track trends, measure rates of deforestation, and evaluate the effectiveness of policy interventions. Governments and conservation groups use these insights to enforce regulations, prioritize protection zones, and allocate resources efficiently. Optical sensors also support reforestation monitoring by tracking the growth of newly planted trees and assessing canopy recovery. The clarity and accessibility of optical imagery make it an essential tool for forest management and environmental oversight.

Radar satellites offer unique advantages for forest monitoring, especially in regions prone to cloud cover or heavy rainfall. Synthetic Aperture Radar (SAR) can penetrate vegetation and detect structural changes in the forest canopy, providing critical information even under challenging weather conditions. Radar data allows scientists to measure biomass, monitor forest degradation, and identify illegal logging operations. Combining radar with optical imagery ensures consistent monitoring, reducing data gaps caused by seasonal changes or atmospheric conditions. Governments, NGOs, and international agencies rely on radar observations to maintain oversight of critical forest regions worldwide. This technology strengthens forest protection efforts and supports global sustainability goals.

Multispectral and hyperspectral imaging enhances the ability to assess forest health and reforestation progress. By measuring reflectance across multiple wavelengths, these sensors detect stress, disease, drought, or nutrient deficiencies in vegetation. Reforestation projects benefit from multispectral monitoring, which confirms tree growth, survival rates, and overall ecosystem recovery. Multispectral data can also guide planting strategies, ensuring that restoration efforts are targeted and effective. By integrating these advanced imaging techniques, satellites provide actionable intelligence for both conservation and sustainable forest management. This technological capability ensures that forest ecosystems are protected and restored efficiently.

Integrating data from multiple satellites provides a comprehensive view of forest dynamics at local, regional, and global scales. Time-series analysis and change detection algorithms allow researchers to track trends, measure deforestation rates, and quantify carbon sequestration over time. Policymakers can use this information to design evidence-based regulations, implement reforestation initiatives, and report progress on international climate goals. Remote sensing also enables cross-border collaboration, as countries can share satellite data to monitor shared forest ecosystems. By providing a holistic and accurate perspective, satellite technology ensures sustainable management and conservation of the world's forests. This integration underscores the critical role of space technology in protecting biodiversity and combating climate change.

Satellite monitoring has been instrumental in tracking deforestation in the Amazon rainforest, one of the most biodiverse regions on Earth. Optical and radar satellites detect illegal logging, agricultural expansion, and infrastructure development, providing near-real-time alerts to authorities. These observations allow governments to enforce environmental regulations and coordinate conservation efforts efficiently. Satellite data also informs international climate programs, as the Amazon is a crucial carbon sink that impacts global climate systems. NGOs and research institutions use the information to advocate for policy change and support sustainable land management initiatives. By providing objective, timely, and precise data, satellites help protect the Amazon's ecosystem and biodiversity.

In Southeast Asia, satellite technology has been used to monitor deforestation caused by palm oil plantations. High-resolution imagery allows authorities and researchers to identify areas where forests are cleared illegally or unsustainably. By tracking forest cover changes over time, satellites provide evidence for sustainable certification programs and guide reforestation initiatives. Local communities and governments can use this information to balance economic development with environmental protection. Monitoring reforestation efforts ensures that newly planted trees survive and contribute to carbon sequestration. Satellite observations have made it possible to enforce regulations and promote sustainable practices across a rapidly changing landscape.

Reforestation projects in Africa demonstrate the effectiveness of satellite technology for ecosystem restoration. Large-scale tree planting programs in countries like Ethiopia and Kenya are monitored using multispectral and radar data. Satellites track tree survival, growth rates, and canopy expansion, ensuring that resources are allocated efficiently. By providing evidence of progress, satellite monitoring attracts further investment and support from international organizations. These observations also inform adaptive management strategies, allowing project managers to adjust techniques based on environmental conditions and tree performance. The integration of satellite data ensures that reforestation initiatives achieve meaningful ecological and social impact.

In Europe, satellite observation supports forest management and wildfire prevention. High-resolution imagery detects changes in vegetation, identifies areas at risk of fire, and monitors post-fire recovery. Countries with extensive forestry sectors use satellite data to plan sustainable harvesting, detect illegal logging, and assess ecosystem health. By tracking both deforestation and reforestation, satellites provide a complete picture of forest dynamics, guiding conservation and economic activities simultaneously. This comprehensive monitoring ensures that forests continue to provide essential ecosystem services, including carbon sequestration, habitat preservation, and soil protection. Satellite data empowers authorities to make informed decisions that balance environmental and economic priorities.

In North America, satellite technology is applied to monitor boreal forests and mitigate the effects of climate change. Satellites detect deforestation, pest outbreaks, and changes in forest density caused by natural and human-induced factors. Reforestation and afforestation projects are tracked to evaluate effectiveness and carbon storage potential. By integrating satellite observations with climate models, scientists assess how forest management practices influence carbon cycles and global warming. These applications ensure that forests contribute to climate mitigation while maintaining biodiversity. Through real-world case studies, satellite monitoring demonstrates its ability to guide effective forest management and environmental stewardship globally.

Satellite monitoring of forests plays a critical role in climate change mitigation by supporting carbon accounting and emissions reduction initiatives. Forests act as natural carbon sinks, absorbing CO_2 from the atmosphere, and satellite data allows scientists to quantify carbon stocks and fluxes accurately. By tracking deforestation, degradation, and reforestation, satellites provide the data needed to estimate emissions and verify compliance with climate agreements. Policymakers can use this information to develop evidence-based strategies for reducing carbon footprints, implementing reforestation programs, and incentivizing sustainable land use. Satellite monitoring also ensures transparency in reporting, enabling international organizations and governments to track progress toward climate goals. Through this integration, space technology directly contributes to global efforts to combat climate change.

Carbon credit programs and sustainable forestry initiatives benefit from satellite observation. By providing verifiable data on tree cover, forest health, and land-use changes, satellites create reliable records for carbon trading and conservation incentives. Countries and companies participating in carbon markets can demonstrate reductions in emissions or increases in sequestration using satellite-derived information. This transparency encourages accountability and promotes investment in reforestation and sustainable forest management projects. Satellite data also identifies areas where interventions are most needed, optimizing the impact of mitigation programs. Integrating remote sensing into carbon credit schemes ensures that climate actions are both measurable and effective.

Satellite monitoring informs national and international forest policy implementation. Governments rely on remote sensing to enforce environmental regulations, design protected areas, and develop sustainable land-use plans. By detecting illegal logging and unauthorized land conversion, satellites support compliance with forestry laws and conservation policies. International agreements, such as REDD+ (Reducing Emissions from Deforestation and Forest Degradation), rely on satellite data to track progress and ensure transparency. Policy decisions based on satellite observations are more precise, targeted, and accountable, enhancing their effectiveness in protecting ecosystems. This integration of technology and governance demonstrates how space-based tools support sustainable development and environmental stewardship.

Integrating forest monitoring with climate models enhances understanding of ecosystem dynamics and future trends. Satellites provide data on forest cover, biomass, and health, which feeds into models that predict carbon storage potential and ecosystem resilience. These predictions help policymakers plan long-term mitigation strategies, including reforestation, afforestation, and conservation initiatives. By understanding the interactions between forests, climate, and human activity, governments can prioritize interventions that maximize environmental and social benefits. Satellite data allows for adaptive management, ensuring that strategies remain effective under changing conditions. The combination of monitoring, modeling, and policy implementation strengthens global climate resilience and forest sustainability.

Long-term satellite observation fosters international cooperation in forest conservation. Shared satellite data enables countries to coordinate efforts to preserve transboundary forests, protect biodiversity, and combat illegal logging. Global monitoring networks facilitate collaboration, capacity building, and knowledge sharing between nations. International organizations rely on this data to track progress, report outcomes, and support funding for sustainable forestry projects. By providing an objective, comprehensive view of forest dynamics, satellites create a common platform for action and accountability. The integration of space technology into climate and forest policy illustrates how collaboration, observation, and science-driven decision-making can protect ecosystems worldwide.

Satellite monitoring of forests has transformed the global approach to biodiversity preservation and climate mitigation. By providing accurate, near-real-time data on deforestation, degradation, and reforestation, satellites empower governments, NGOs, and international organizations to make informed decisions that protect ecosystems. These insights help maintain habitat connectivity, safeguard endangered species, and ensure the survival of critical biodiversity hotspots. Monitoring reforestation progress ensures that restoration initiatives achieve measurable ecological outcomes and contribute to carbon sequestration. By tracking trends over time, satellite data allows for adaptive management strategies, enhancing the effectiveness of conservation and restoration programs. The comprehensive perspective offered by satellites ensures that forests continue to provide essential ecological, social, and economic benefits.

Satellites also enable accountability and transparency in forest management on a global scale. By independently verifying deforestation and reforestation activities, satellite observations hold governments, corporations, and communities responsible for their environmental impact. International climate initiatives, such as REDD+ and global carbon credit programs, rely on this data to validate progress and allocate resources effectively. Transparent monitoring encourages compliance with environmental regulations and fosters trust among stakeholders. It also incentivizes investment in sustainable forest management and reforestation projects that deliver measurable benefits. Space-based observation, therefore, supports both environmental stewardship and global cooperation.

Forest monitoring from orbit contributes directly to climate change mitigation efforts. By quantifying carbon storage, emissions from deforestation, and sequestration from reforestation, satellites provide critical input for climate models and policy decisions. This information guides emission reduction strategies, reforestation planning, and sustainable land-use practices. Governments and international organizations can use satellite data to target interventions where they will have the greatest impact. By integrating forest monitoring into climate policies, satellites help ensure that forests continue to function as vital carbon sinks. Their role demonstrates the power of space technology to address one of the most pressing environmental challenges of our time.

In addition to climate benefits, satellite forest monitoring supports sustainable development and community livelihoods. Accurate mapping of forest resources enables local communities to manage timber, non-timber forest products, and ecosystem services sustainably. It also informs disaster risk management, such as identifying areas prone to landslides or floods due to deforestation. By combining conservation, economic development, and disaster preparedness, satellites help create resilient and sustainable communities. This multidimensional approach highlights how space technology can balance ecological, social, and economic priorities. Through forest monitoring, satellites provide a foundation for sustainable development on a global scale.

Ultimately, mapping deforestation and reforestation from orbit illustrates the transformative impact of space technology on Earth's ecosystems. Satellites provide the knowledge, precision, and transparency needed to protect biodiversity, mitigate climate change, and manage forests sustainably. They enable governments, organizations, and communities to take evidence-based action that preserves natural resources for current and future generations. By integrating monitoring with policy, climate strategies, and international cooperation, satellites ensure that forest ecosystems continue to thrive. Space technology, therefore, is an indispensable tool for safeguarding Earth's forests, supporting sustainable development, and securing a healthier planet for all living beings.

Chapter . 9

Tracking Ocean Health to Protect Marine Life

Using satellite data to monitor ocean temperatures, currents, and pollution for sustainable fisheries.

Satellite technology has revolutionized the study and management of ocean health, enabling scientists to observe marine ecosystems on a global scale. Satellites monitor sea surface temperature, ocean color, currents, and biological activity, providing crucial data for assessing the health of oceans and their biodiversity. These observations inform fisheries management, climate studies, and conservation efforts, ensuring sustainable use of marine resources. By providing continuous, high-resolution data, satellites allow researchers to detect subtle changes that could threaten marine life. Governments, international organizations, and NGOs rely on this information to guide policy, enforce regulations, and protect vulnerable ecosystems. Space-based ocean monitoring therefore plays a critical role in maintaining the balance and sustainability of marine environments.

Monitoring sea surface temperature is essential for understanding marine ecosystem health. Satellites detect temperature variations across vast ocean areas, revealing patterns related to climate phenomena such as El Niño and La Niña. These temperature shifts influence fish migration, coral bleaching, and the productivity of marine food webs. By tracking changes over time, satellites help scientists predict impacts on fisheries and coastal communities. Governments and fisheries organizations can use this information to manage stocks, regulate catches, and prevent overfishing. Understanding sea surface temperature also supports climate models that predict future oceanic conditions and their ecological consequences.

Satellite observation of ocean color provides insights into phytoplankton abundance and primary productivity. Phytoplankton form the base of the marine food chain and are essential for carbon sequestration. Changes in ocean color indicate shifts in nutrient levels, plankton populations, and potential harmful algal blooms. These data help researchers monitor ecosystem health, detect environmental stress, and assess the effects of climate change on marine biodiversity. By providing near-real-time information, satellites enable timely interventions to protect fisheries and marine habitats. Ocean color monitoring ensures that marine resources are managed sustainably and that ecological balance is maintained.

Currents, tides, and ocean circulation patterns are also monitored via satellites. These physical processes influence nutrient distribution, migration of marine species, and overall ocean health. Satellite data allows scientists to map currents, track eddies, and model the movement of pollutants or invasive species. Understanding circulation patterns helps in designing marine protected areas, managing fisheries, and predicting the impact of climate change on ocean systems. Accurate monitoring of ocean currents also supports disaster preparedness, such as forecasting the spread of oil spills or the effects of hurricanes. By observing these dynamic processes, satellites provide essential guidance for sustainable ocean management.

Pollution monitoring is another critical application of satellite technology for ocean health. Satellites detect oil spills, plastic accumulation, sediment runoff, and chemical contamination over large areas. Remote sensing allows for rapid identification of pollution sources and assessment of environmental impacts. Governments and conservation organizations can respond more effectively, coordinating cleanup efforts and implementing preventive measures. Satellite monitoring also supports international agreements on marine pollution control, ensuring compliance and accountability. By providing timely, accurate, and comprehensive data, satellites help protect marine ecosystems from human-induced threats and maintain the productivity of the oceans.

Satellite technology plays a vital role in monitoring marine biodiversity and ecosystem dynamics. By observing ocean color, temperature, and chlorophyll concentrations, satellites can detect the distribution and abundance of phytoplankton, which form the foundation of the marine food web. This information helps scientists understand the productivity of fisheries and the health of marine species from small invertebrates to large predators. Tracking biodiversity patterns allows policymakers to implement conservation strategies, protect endangered species, and maintain ecosystem balance. Satellite monitoring also supports marine spatial planning, guiding the designation of protected areas and sustainable fishing zones. By linking ecosystem observations with human activity, satellites ensure that resource management is both effective and ecologically responsible.

Harmful algal blooms, which can devastate marine life and human health, are also detected using satellite data. Changes in ocean color and chlorophyll concentrations signal the onset of these blooms, enabling early intervention. Satellite observations allow authorities to warn fisheries, aquaculture operations, and coastal communities, minimizing economic losses and health risks. By tracking the spread of blooms, satellites guide research into environmental triggers and long-term mitigation strategies. Early detection and monitoring help preserve marine biodiversity and prevent the collapse of local fisheries. This capability demonstrates how space-based observation directly contributes to safeguarding ocean ecosystems.

Satellites also provide data for sustainable fisheries management. Monitoring ocean temperature, currents, and productivity helps determine the location and abundance of commercially important fish species. This information allows fisheries to plan harvesting schedules, avoid overfishing, and maintain stock sustainability. International regulatory bodies rely on satellite observations to enforce quotas, monitor compliance, and protect endangered species. By combining satellite data with on-site monitoring, fisheries management becomes more precise, adaptive, and sustainable. Satellite-assisted decision-making ensures that marine resources can continue to provide food, employment, and economic benefits for future generations.

Ocean health monitoring through satellites also supports climate change research and mitigation strategies. Changes in sea surface temperature, currents, and phytoplankton productivity influence global carbon cycles and climate regulation. Satellites provide essential data for modeling the interactions between ocean ecosystems and the atmosphere. This information helps scientists predict the impacts of climate change on marine biodiversity and fisheries productivity. Policymakers can use these insights to develop adaptive strategies, such as managing marine protected areas or adjusting fishing practices. Satellite observation thus links ocean conservation with broader climate action initiatives, ensuring integrated and informed decision-making.

By integrating satellite data with predictive models, ocean health can be forecasted and managed proactively. Early detection of stressors, including warming events, harmful algal blooms, and pollution, enables timely responses that reduce ecological and economic damage. These predictive capabilities support emergency response planning, such as mobilizing resources to address oil spills or responding to sudden shifts in fish populations. Real-time monitoring combined with predictive modeling ensures that interventions are effective and timely. Satellite-enabled forecasting strengthens the resilience of marine ecosystems and the communities that depend on them. By providing actionable insights, satellites help protect ocean health while supporting sustainable development goals.

Satellite monitoring has been essential in managing the fisheries of the North Atlantic. By tracking sea surface temperature, chlorophyll levels, and ocean currents, satellites help determine the location of key fish populations, including cod, herring, and mackerel. This data allows fisheries managers to optimize harvests while preventing overfishing and ensuring the long-term sustainability of stocks. High-resolution imagery also identifies areas affected by environmental stressors, such as pollution or temperature anomalies, that could reduce fish populations. By integrating satellite data with local observations, authorities can develop adaptive management plans that protect both ecosystems and livelihoods. The North Atlantic example demonstrates how space-based observation enables sustainable fisheries management on a large scale.

In the Pacific Ocean, satellites support coral reef protection and conservation initiatives. Coral reefs are highly sensitive to temperature fluctuations, ocean acidification, and pollution, making them vulnerable to climate change and human activity. Satellites monitor sea surface temperature, sediment plumes, and coastal land-use changes to identify stress events and bleaching risks. Early detection enables targeted interventions, such as marine protected area enforcement, pollution mitigation, and coral restoration projects. By tracking the recovery of reefs over time, satellites provide objective data to assess the effectiveness of conservation efforts. This continuous monitoring helps preserve critical habitats that support marine biodiversity and protect coastal communities from storm surges.

Satellite technology has also played a key role in monitoring ocean pollution, including oil spills and plastic accumulation. High-resolution imagery detects oil slicks, floating debris, and sediment runoff from rivers, allowing rapid response and containment. Satellites track the movement of pollutants across vast ocean areas, helping authorities deploy cleanup efforts strategically and efficiently. Remote sensing also identifies sources of pollution, supporting enforcement of environmental regulations and international agreements. This data informs public awareness campaigns, scientific research, and policy decisions related to marine protection. By detecting pollution events early, satellites help minimize environmental damage and preserve marine life.

In Southeast Asia, satellite monitoring supports sustainable management of shrimp and tuna fisheries. Remote sensing identifies temperature gradients, chlorophyll-rich zones, and ocean currents that influence the distribution of commercially important species. Fisheries use this information to optimize harvesting, reduce bycatch, and minimize ecological impacts. Satellite data also informs regional management plans, ensuring that fishing practices are aligned with conservation objectives. By providing near-real-time data, satellites enable adaptive strategies that maintain productivity while protecting marine biodiversity. This approach demonstrates how space technology can balance economic and environmental priorities in highly productive ocean regions.

Satellite monitoring also supports global research networks studying ocean health and climate change. International initiatives such as the Global Ocean Observing System (GOOS) rely on satellite data to track ocean conditions, monitor ecosystems, and assess the impacts of human activity. By sharing satellite observations, researchers can identify trends, detect anomalies, and develop predictive models for ocean health. These collaborations enhance scientific understanding, inform policy, and guide conservation programs worldwide. Space-based ocean monitoring provides a shared knowledge platform that benefits governments, NGOs, communities, and industries alike. Through these applications, satellites help ensure the long-term sustainability of the world's oceans.

Satellite data integration with climate models enhances our understanding of how oceans influence global climate systems. Sea surface temperature, currents, and chlorophyll concentrations are incorporated into predictive models to assess carbon cycles, heat transport, and ecosystem responses. These models allow scientists to anticipate the effects of climate change on fisheries, coral reefs, and marine biodiversity. By linking satellite observations with oceanographic simulations, researchers can forecast changes in productivity, species distribution, and ecosystem health. Policymakers rely on these insights to develop strategies for climate adaptation, marine conservation, and sustainable resource management. The integration of satellite data and climate models ensures informed decision-making that balances environmental, social, and economic priorities.

Disaster preparedness also benefits from satellite ocean monitoring. Hurricanes, cyclones, and storm surges are influenced by sea surface temperature, ocean currents, and wind patterns, all of which are measured by satellites. Early detection and monitoring allow authorities to issue warnings, coordinate evacuations, and prepare emergency response plans. Satellite data also informs the management of fisheries and coastal ecosystems in the aftermath of storms, guiding recovery and minimizing economic and ecological damage. By providing continuous observation and predictive capabilities, satellites improve the resilience of both natural systems and human communities. The combination of real-time monitoring and modeling ensures proactive disaster management in marine and coastal areas.

Global policy and international cooperation are strengthened through shared satellite observations of ocean health. International organizations, such as the United Nations and regional fisheries management bodies, use satellite data to set sustainable fishing quotas and establish marine protected areas. Cross-border collaboration ensures that migratory species, shared fisheries, and oceanic ecosystems are managed collectively and sustainably. Transparent and standardized satellite data fosters trust among nations, supports compliance with environmental agreements, and guides conservation initiatives. By providing a common, verifiable source of information, satellites facilitate global action on marine sustainability. This cooperation ensures that the benefits of space technology extend to oceans worldwide, regardless of a nation's individual monitoring capacity.

Monitoring ocean pollution using satellites informs both local and global mitigation strategies. Oil spills, chemical contaminants, and plastic waste are tracked in near-real time, enabling rapid response and cleanup efforts. Satellite data also identifies pollution sources, guiding regulatory enforcement and industrial compliance. Pollution monitoring contributes to international agreements and standards, ensuring that marine ecosystems are protected and sustainable practices are upheld. By combining satellite observation with policy and enforcement, governments can prevent long-term ecological damage while supporting economic activities that depend on healthy oceans. This proactive approach demonstrates the critical role of satellites in safeguarding marine environments.

Satellite monitoring has fundamentally transformed our ability to protect marine biodiversity and sustain ocean resources. By tracking sea surface temperature, currents, chlorophyll concentrations, and pollution, satellites provide the data necessary to manage fisheries, conserve habitats, and maintain ecosystem balance. These observations help policymakers, scientists, and conservation organizations make informed decisions that protect marine life and ensure the long-term sustainability of ocean resources. Continuous monitoring allows for early detection of stressors, from harmful algal blooms to coral bleaching events, enabling timely interventions that prevent ecological collapse. Satellite data also supports reforestation of mangroves and restoration of coastal habitats, contributing to both biodiversity preservation and climate mitigation. Through these capabilities, satellites have become essential tools for maintaining the health and resilience of the world's oceans.

Sustainable fisheries management benefits greatly from satellite technology. Observing fish distributions, migration patterns, and environmental conditions allows authorities to implement quotas, reduce overfishing, and protect endangered species. By integrating satellite data with on-the-ground monitoring, fisheries are able to balance economic productivity with ecological sustainability. Communities that rely on fishing for food and income gain resilience from these evidence-based management practices. Satellite monitoring ensures that global fisheries remain productive while preserving the ecological integrity of marine ecosystems. This dual benefit highlights the transformative impact of space technology on both human livelihoods and environmental conservation.

Pollution detection and management are also enhanced through satellites. Oil spills, chemical discharges, and plastic debris are tracked over vast ocean areas, providing timely information for containment and cleanup operations. Satellite data allows governments and organizations to identify pollution sources, enforce regulations, and measure compliance with international agreements. Monitoring pollution contributes directly to protecting marine life, maintaining ecosystem services, and safeguarding the health of coastal communities. Satellites provide a transparent, verifiable method for evaluating the effectiveness of environmental policies and interventions. By reducing ecological damage, satellite monitoring supports sustainable ocean management worldwide.

Satellite observations are critical for understanding the impact of climate change on ocean systems. Changes in temperature, currents, and productivity influence marine species distribution, coral reef health, and carbon cycling. By providing long-term datasets, satellites allow scientists to model the effects of climate change on marine ecosystems and forecast potential risks. Policymakers can use these insights to develop adaptation strategies, such as adjusting fishing practices, expanding marine protected areas, or restoring coastal habitats. The integration of satellite data with climate research ensures that marine conservation and resource management are proactive rather than reactive. Space technology, therefore, underpins global efforts to maintain resilient oceans in the face of a changing climate.

Ultimately, tracking ocean health from orbit demonstrates the indispensable role of satellite technology in sustaining marine ecosystems. Satellites provide comprehensive, accurate, and timely data that informs conservation, fisheries management, pollution control, and climate adaptation. By enabling proactive decision-making, evidence-based policy, and international cooperation, satellites ensure that oceans continue to support biodiversity, human livelihoods, and global environmental stability. The knowledge gained from space-based monitoring guides sustainable management practices that protect marine life for current and future generations. Through observation, prediction, and policy support, satellites have become a cornerstone of ocean stewardship and the preservation of the planet's most vital ecosystems.

Chapter . 10

Managing Water Resources with Space-Based Insights

Employing satellite monitoring to optimize freshwater usage and prevent shortages.

Water is a critical resource for human survival, agriculture, industry, and ecosystem health, yet managing it sustainably has become increasingly challenging. Satellite technology offers a powerful tool to monitor and manage water resources on a global scale. Satellites measure surface water levels, precipitation, soil moisture, snowpack, and groundwater storage, providing comprehensive data for decision-makers. This information helps governments, farmers, and communities optimize water allocation, plan for droughts, and prevent shortages. By offering near-real-time insights, satellites enable proactive management rather than reactive responses to water crises. Space-based water monitoring ensures that freshwater resources are used efficiently, equitably, and sustainably.

Monitoring surface water bodies, such as lakes, rivers, and reservoirs, is essential for effective water management. Satellites provide accurate measurements of water volume, flow rates, and seasonal variations. This data allows water authorities to plan irrigation schedules, manage reservoir levels, and ensure reliable water supply for cities and agriculture. By detecting changes in water storage, satellite observations help predict floods or shortages before they become crises. Satellite monitoring also informs hydropower management, balancing energy production with ecological needs. The ability to track surface water globally enhances preparedness, resilience, and sustainability in water management practices.

Satellite measurements of precipitation and snowpack are vital for predicting water availability. Satellites detect rainfall patterns, storm intensity, and snowfall accumulation across large areas, including regions that lack ground-based monitoring. This information informs river flow forecasts, reservoir management, and flood prevention strategies. Accurate precipitation data also supports agricultural planning, ensuring crops receive adequate water while reducing waste. Snowpack monitoring is especially important in regions where melting snow provides a significant portion of freshwater supply. By providing comprehensive observations, satellites enable efficient and reliable management of water resources across diverse climates and landscapes.

Soil moisture monitoring using satellites improves agricultural water efficiency and drought management. By measuring water content in the soil, satellites help farmers determine when and how much to irrigate, reducing waste and conserving freshwater. Soil moisture data also informs drought monitoring and early warning systems, allowing authorities to implement mitigation strategies before crop failures occur. Additionally, soil moisture observations contribute to flood prediction by identifying areas of saturation that may be at risk of runoff or landslides. Integrating soil moisture data with precipitation and surface water measurements creates a comprehensive understanding of regional water availability. Satellites thus provide critical intelligence for both agricultural productivity and sustainable water management.

Groundwater assessment is enhanced through space-based techniques such as gravimetry and remote sensing. Satellites like GRACE (Gravity Recovery and Climate Experiment) detect changes in Earth's gravity field caused by variations in groundwater storage. This information enables scientists to track aquifer depletion, recharge rates, and long-term water sustainability. By understanding groundwater dynamics, governments can manage extraction, prevent overuse, and protect communities reliant on underground water sources. Satellite-based groundwater monitoring also informs regional water policies, helping prevent conflicts over shared water resources. Space technology, therefore, provides a unique and comprehensive perspective on all components of the water cycle, ensuring that freshwater resources are managed effectively for human and environmental needs.

Satellite data is transforming agricultural water management by providing precise information on soil moisture, precipitation, and irrigation needs. Farmers can use this information to schedule watering efficiently, minimizing waste while ensuring optimal crop growth. Satellite observations also allow for monitoring of water use across large agricultural regions, helping governments enforce sustainable practices. By reducing over-irrigation, satellite-guided farming conserves freshwater resources, lowers costs, and mitigates environmental impacts such as soil salinization and nutrient runoff. Data from satellites also supports drought planning, enabling proactive adjustments to cropping patterns and irrigation strategies. Integrating space-based insights into agriculture ensures that water is used efficiently, supporting both food security and environmental sustainability.

Urban water supply management benefits significantly from satellite monitoring. Cities can track reservoir levels, river flows, and water quality to ensure a consistent and safe supply for residents. Real-time satellite data helps water utilities predict shortages, optimize distribution, and allocate resources efficiently during periods of high demand or drought. Satellite observations also aid in monitoring water quality by detecting sediment, algae blooms, and pollution events in lakes, rivers, and reservoirs. This information supports timely interventions, such as treatment adjustments or public advisories, to protect human health. By integrating satellite data into urban water management systems, municipalities can improve resilience and reliability while reducing environmental stress.

Drought monitoring and early warning systems rely heavily on satellite observations. By combining precipitation data, soil moisture measurements, and surface water levels, satellites detect emerging drought conditions before they become severe. Governments and farmers can use this information to implement mitigation strategies, including rationing, water recycling, or emergency supply planning. Early warning allows for proactive responses, reducing crop losses, economic damage, and human suffering. Satellites also help track the effectiveness of drought management strategies, providing feedback to improve future planning. Space-based monitoring thus enhances preparedness and reduces the risks associated with water scarcity.

Flood prevention and management are similarly enhanced through satellite observation. Satellites track rainfall, snowmelt, river levels, and soil saturation to forecast potential flooding events. This data enables early warnings, evacuation planning, and allocation of emergency resources, reducing human and economic losses. Satellite imagery also helps map flood-prone areas, guiding infrastructure development, zoning, and disaster preparedness plans. Post-flood monitoring allows authorities to assess damage, plan recovery efforts, and improve future flood mitigation strategies. By providing continuous, accurate, and comprehensive data, satellites enhance flood resilience and protect both people and ecosystems.

Integrating water resource data from multiple satellite sources creates a holistic understanding of the hydrological cycle. Surface water, groundwater, soil moisture, precipitation, and snowpack measurements combined provide a complete picture of water availability and stress points. Governments, agencies, and researchers can analyze this integrated data to plan sustainable water use, balance competing demands, and prioritize conservation efforts. This comprehensive approach ensures that freshwater resources are managed equitably, efficiently, and resiliently. By linking space-based observations with local water management practices, satellites enable evidence-based policies that support long-term water security. Space technology thus transforms water management into a proactive, precise, and sustainable endeavor.

Satellite technology has been transformative in managing water resources in arid regions, where scarcity poses major challenges for agriculture and communities. In places like the Middle East and North Africa, satellites monitor groundwater levels, soil moisture, and surface water availability to guide efficient irrigation practices. By identifying regions where water resources are critically low, authorities can implement rationing, water recycling, and sustainable management strategies. Satellite data also helps track desalination outputs and reservoir storage, ensuring reliable water supply despite harsh climatic conditions. The integration of satellite observations with local water management practices allows these regions to maximize limited resources sustainably. Through space-based insights, arid regions can maintain agricultural productivity and secure water for urban populations.

In India, satellite monitoring supports large-scale agricultural water management in river basins like the Ganges. Satellites track rainfall patterns, soil moisture, and reservoir levels to guide irrigation scheduling and flood control measures. Real-time data allows farmers to optimize water usage, reducing waste while protecting crop yields. Government agencies use satellite observations to implement drought mitigation strategies and manage seasonal water distribution effectively. By combining satellite data with hydrological modeling, India can anticipate potential water shortages and respond proactively. This integrated approach demonstrates how space-based monitoring improves both agricultural productivity and water security in densely populated regions.

Urban water management in megacities like Los Angeles and São Paulo relies heavily on satellite observations for efficient operation. Satellites provide data on reservoir levels, river inflows, and urban water demand, enabling city planners to allocate water resources effectively. Early detection of droughts, water stress, or pollution allows municipalities to take preventive action, protecting residents and industries. Satellite monitoring also supports infrastructure planning, such as expanding reservoirs or optimizing pipeline networks. By integrating space-based insights with local water management systems, cities can maintain resilience against growing population pressures and climate variability. Satellites thus enhance urban water security while supporting sustainable development.

In South America, satellites are used to monitor the Amazon Basin's rivers and wetlands to ensure sustainable water use and ecosystem health. Satellite imagery tracks seasonal flooding, river discharge, and sediment transport, providing critical data for water resource management. This information helps balance the needs of agriculture, hydropower, and biodiversity conservation. By detecting early signs of drought or abnormal water levels, authorities can implement adaptive management strategies to prevent ecological and economic damage. Satellite monitoring also supports long-term research on hydrological cycles, climate change impacts, and water-dependent ecosystems. Space technology thus allows for sustainable water management across vast and ecologically sensitive regions.

Satellite monitoring also supports international cooperation in transboundary water management. Shared rivers, aquifers, and watersheds often span multiple countries, making coordinated management essential. Satellites provide objective, verifiable data on water availability, usage, and changes in hydrology. This shared knowledge enables neighboring countries to negotiate resource allocation, prevent conflicts, and implement joint conservation measures. International organizations rely on satellite observations to assess compliance with agreements and support sustainable water policies. By enabling transparency and collaboration, satellites ensure that water resources are managed fairly and sustainably across borders.

Satellite observations are essential for integrating water management with climate adaptation strategies. By providing accurate data on precipitation, snowpack, soil moisture, and groundwater, satellites help governments and communities anticipate and respond to changing climate conditions. This information allows planners to develop adaptive strategies, such as drought-resistant crop selection, reservoir management, and urban water infrastructure improvements. Early detection of climate-related water stress enables proactive mitigation measures that reduce economic losses and protect vulnerable populations. Satellites also provide long-term data that informs modeling of future water availability under different climate scenarios. By linking observation with adaptation planning, space technology ensures water security in an era of climate uncertainty.

Disaster risk reduction benefits significantly from satellite monitoring of water resources. Floods, droughts, and water contamination events are detected early through satellite data, enabling timely warnings and emergency response planning. Predictive models using satellite observations allow authorities to assess vulnerability, prioritize interventions, and deploy resources effectively. By monitoring rivers, reservoirs, and precipitation patterns, satellites help minimize the impact of natural disasters on communities, agriculture, and critical infrastructure. Post-disaster assessment is also facilitated by satellite imagery, providing a clear picture of damage and guiding recovery and reconstruction. This proactive approach to disaster management enhances resilience and reduces the human and economic costs of water-related hazards.

Policy frameworks for sustainable water management increasingly rely on space-based insights. Governments use satellite data to implement evidence-based regulations, monitor compliance with water laws, and plan equitable water distribution. International treaties on transboundary rivers and shared aquifers are strengthened through satellite monitoring, providing transparent and verifiable data for negotiations. Policy decisions based on satellite observations ensure that water allocation balances social, economic, and ecological needs. By incorporating space technology into governance, policymakers can make informed, data-driven choices that promote sustainable water use. Satellite data thus becomes a critical tool for both national and international water governance.

Integrating satellite water data with local monitoring systems enhances precision and reliability. Ground-based sensors, weather stations, and hydrological measurements complement satellite observations, providing high-resolution insights at regional and local levels. This integration allows water managers to develop detailed allocation plans, optimize irrigation, and respond to emerging water stress conditions. It also improves forecasting for agriculture, urban supply, and environmental conservation. By combining multiple data sources, stakeholders gain a comprehensive understanding of the water cycle, ensuring more effective and resilient management. Space-based technology therefore acts as a bridge between global monitoring and local decision-making.

Satellite monitoring of water resources has revolutionized the management of freshwater systems worldwide. By providing accurate, real-time information on surface water, groundwater, soil moisture, and precipitation, satellites enable proactive and sustainable decision-making. Agriculture benefits from precise irrigation planning, minimizing water waste and maximizing crop productivity. Urban centers rely on satellite data to maintain reliable water supply, detect pollution, and respond effectively to shortages. Drought and flood management are enhanced through early warning systems that allow communities to prepare and reduce economic and human losses. Space-based water monitoring ensures that freshwater resources are used efficiently, equitably, and sustainably.

The integration of satellite data into agricultural planning improves food security and environmental sustainability. Farmers can adjust irrigation schedules, select drought-resistant crops, and monitor soil health based on satellite observations. Governments and agencies can track water usage across regions, enforce sustainable practices, and optimize resource allocation. By reducing overuse and enhancing efficiency, satellite-guided agriculture conserves water, protects ecosystems, and supports economic stability. Satellite monitoring bridges the gap between water availability and agricultural demand, ensuring resilience in the face of climate variability. This technology demonstrates how space-based insights directly benefit human livelihoods.

Urban water management is strengthened through satellite-enabled monitoring of reservoirs, rivers, and supply networks. Real-time observations allow cities to anticipate shortages, detect contamination, and allocate resources effectively. By combining satellite data with local infrastructure monitoring, municipalities can maintain safe and reliable water supply even during periods of drought or extreme weather. Satellite observations also support long-term planning for urban expansion and climate adaptation. This integrated approach ensures that cities remain resilient, sustainable, and capable of meeting growing water demands. Space technology, therefore, plays a central role in urban water security.

Satellite monitoring also enhances disaster preparedness and climate adaptation related to water resources. Early detection of droughts, floods, and abnormal hydrological events enables proactive mitigation, minimizing losses and protecting vulnerable populations. Long-term satellite datasets inform climate models and water availability forecasts, allowing policymakers to develop adaptive strategies for agriculture, urban management, and ecosystem conservation. By providing timely, accurate, and comprehensive data, satellites strengthen the capacity of communities and governments to respond to emerging water challenges. This proactive approach transforms water management from reactive crisis response to strategic planning.

Ultimately, managing water resources from space illustrates the transformative impact of satellite technology on Earth's most vital resource. Satellites provide the knowledge, precision, and foresight needed to optimize water use, safeguard ecosystems, and secure freshwater availability for current and future generations. By supporting agriculture, urban supply, disaster preparedness, and global cooperation, satellite monitoring ensures that water management is sustainable, equitable, and resilient. Space technology has become indispensable in addressing the challenges of water scarcity, climate change, and growing human demand. Through observation, analysis, and informed decision-making, satellites safeguard one of humanity's most essential resources while promoting global sustainability.

Part III: Energy, Agriculture, and Resource Management

Chapter . 11

Optimizing Solar Energy with Space Research

Translating solar innovations from space missions to improve renewable energy on Earth.

Space research has long driven innovations in solar energy technology, offering solutions that extend far beyond satellites and spacecraft. Solar panels designed for space missions must operate efficiently in extreme conditions, including temperature fluctuations, radiation exposure, and microgravity. Engineers and scientists apply these innovations to terrestrial solar technology, improving efficiency, durability, and energy output. By translating lessons learned from orbital energy systems, solar power on Earth becomes more reliable and cost-effective. Research in space also inspires new materials, tracking systems, and energy storage solutions that maximize solar energy capture. Ultimately, insights from space research help accelerate the global transition to clean, renewable energy.

High-efficiency photovoltaic cells developed for spacecraft have revolutionized terrestrial solar panels. Space-grade cells capture more sunlight and convert it to electricity with minimal energy loss, even under challenging conditions. These advancements have been adapted for solar farms, residential rooftops, and portable energy solutions on Earth. Research into lightweight and flexible solar materials from space applications also opens opportunities for urban and mobile installations. By adopting these innovations, renewable energy systems become more adaptable, sustainable, and efficient. Space research continues to push the boundaries of what is possible in solar technology, providing tangible benefits for global energy production.

Solar tracking systems, originally designed for spacecraft orientation, enhance energy efficiency on Earth. These systems adjust the angle of solar panels throughout the day to follow the sun, maximizing energy capture. Implementing tracking technology in solar farms increases electricity generation without requiring additional land or panels. Space research also informs algorithms and automation techniques for optimal panel positioning. By integrating tracking systems with predictive weather and sunlight data, solar energy can be optimized for both production and grid stability. This direct application of space technology significantly improves the return on investment for renewable energy projects.

Energy storage innovations from space research further improve the reliability of solar power on Earth. Space missions require compact, efficient batteries capable of storing energy for long periods and operating under harsh conditions. Adapting these technologies enables better energy storage solutions for solar grids, reducing dependency on fossil fuels and providing stable power even during cloudy periods or at night. Advanced battery materials, thermal management systems, and smart energy distribution methods developed for space applications now support terrestrial energy resilience. By enhancing storage capabilities, satellites and spacecraft indirectly improve the feasibility of widespread solar energy adoption.

Beyond hardware, space research contributes to solar energy planning and optimization through remote sensing and modeling. Satellites monitor cloud cover, solar irradiance, and environmental conditions, providing data for efficient solar farm placement and operation. These insights help energy planners select optimal locations, predict power output, and integrate solar energy into broader grids. By leveraging satellite observations and modeling techniques, solar energy deployment can be maximized for efficiency, cost-effectiveness, and environmental benefit. Space research thus not only drives technological innovation but also informs practical strategies for sustainable energy expansion on Earth.

Space-derived solar technology has already made a tangible impact on terrestrial energy systems. Large-scale solar farms benefit from high-efficiency panels, tracking systems, and durable materials initially developed for satellites. By implementing these innovations, solar farms generate more electricity per unit area while reducing maintenance costs and energy losses. In urban environments, flexible and lightweight solar panels inspired by space applications can be integrated into rooftops, building facades, and even windows. This allows cities to harness solar energy without requiring additional land, supporting sustainable urban development. Portable solar solutions, such as foldable panels for disaster relief or remote communities, also draw directly from space research innovations. These applications demonstrate how lessons from orbital technology translate into practical, Earth-based energy solutions.

High-altitude solar research platforms provide unique insights into optimizing terrestrial energy systems. Balloons, drones, and stratospheric solar simulators, inspired by space deployment methods, test solar performance under varying atmospheric conditions. These experiments help engineers improve panel efficiency, heat management, and energy conversion. The resulting data informs the design of terrestrial solar arrays, allowing for better performance under real-world environmental conditions. By learning from space-inspired platforms, solar energy systems can operate more efficiently and reliably across diverse climates and geographies. This continuous feedback loop between space research and Earth applications drives incremental improvements in renewable energy deployment.

Satellite-based solar monitoring plays a critical role in planning and optimizing energy production on Earth. Observations of cloud cover, solar irradiance, and atmospheric conditions enable precise predictions of solar power availability. Energy grids can use this information to balance supply, manage storage, and ensure reliability for residential, commercial, and industrial consumers.

Forecasting energy production based on satellite data also allows for integration with other renewable sources, such as wind or hydroelectric power. This strategic planning reduces dependency on fossil fuels and enhances energy resilience. Space-based observation thus supports not only technology development but also operational efficiency and grid management.

Urban solar installations benefit from innovations inspired by spacecraft energy systems. Lightweight, flexible, and durable panels enable the integration of solar technology into building surfaces, transportation networks, and public infrastructure. Advanced materials developed for satellites resist environmental wear, corrosion, and temperature fluctuations, extending the lifespan of urban solar arrays. By adapting space-grade components for city environments, solar energy becomes more scalable, aesthetically compatible, and cost-effective. Cities can therefore maximize renewable energy capture while minimizing disruption to existing infrastructure. Space research provides the engineering breakthroughs necessary to make urban solar power a practical, sustainable solution.

Portable solar solutions for remote or emergency applications draw heavily from space technology. Foldable, lightweight, and high-efficiency panels developed for spacecraft are ideal for disaster relief, off-grid communities, and exploration missions. These systems provide reliable electricity in areas without traditional grid access, supporting essential services such as healthcare, communication, and water purification. Emergency response teams use portable solar technology to maintain operations during natural disasters or humanitarian crises. By transferring lessons learned from space missions to portable energy solutions, satellites indirectly enhance resilience, sustainability, and quality of life on Earth. Space research thus empowers energy access in some of the world's most challenging environments.

Integrating solar energy into national and regional grids is greatly improved by space-derived technology and data. High-efficiency panels, tracking systems, and predictive monitoring allow energy providers to maximize electricity generation while balancing demand and supply. Satellite observations of solar irradiance, cloud cover, and atmospheric conditions provide critical inputs for grid management, ensuring consistent and reliable power. These insights allow for strategic planning, reducing dependence on fossil fuels and minimizing disruptions caused by variable solar output. By combining technological innovations with satellite-based forecasting, energy grids can efficiently incorporate higher percentages of renewable energy. This integration demonstrates how space research directly informs and strengthens terrestrial energy infrastructure.

Advanced energy storage solutions developed for spacecraft are revolutionizing solar energy reliability on Earth. Satellites rely on highly efficient batteries and energy management systems to maintain continuous power during periods of darkness or limited sunlight. Adapting these technologies enables terrestrial solar systems to store energy more effectively, providing electricity at night or during cloudy conditions. Innovations in battery chemistry, thermal regulation, and smart distribution management are applied to grid-scale, residential, and off-grid storage systems. Reliable storage ensures uninterrupted energy supply and facilitates the integration of solar power into broader energy networks. By leveraging space-tested storage solutions, solar energy becomes a viable and resilient alternative to traditional energy sources.

Predictive modeling inspired by space missions improves the efficiency and deployment of solar energy projects. Space research relies on simulations to optimize satellite power systems, which has translated into advanced algorithms for predicting solar panel output on Earth. These models consider weather patterns, shading, seasonal changes, and regional irradiance variations to estimate electricity generation accurately. Utilities and solar farm operators can use these predictions to optimize performance, manage storage, and coordinate energy distribution. Predictive modeling also supports investment decisions and policy development, ensuring that solar energy projects are economically viable and environmentally sustainable. Space-based insights, therefore, enhance planning, efficiency, and strategic implementation of renewable energy systems.

Lessons from spacecraft engineering inspire innovations in materials, design, and system efficiency for solar installations. Lightweight, flexible, and radiation-resistant materials used in satellites improve durability and efficiency in terrestrial solar panels. Space-inspired design strategies, such as modular arrays and foldable configurations, allow for rapid deployment and maintenance in diverse environments. These innovations increase the lifespan of solar installations, reduce costs, and expand opportunities for adoption in urban, rural, and off-grid settings. By applying engineering principles tested in space, renewable energy systems on Earth benefit from higher performance, reliability, and adaptability. Space research thus drives continuous improvement in solar technology applications worldwide.

International collaboration in space-based solar research accelerates global renewable energy adoption. Space agencies, universities, and private companies share data, technology, and innovations that improve solar efficiency, storage, and deployment. Lessons learned from solar-powered spacecraft missions inform best practices, equipment standards, and operational strategies on Earth. Collaborative research also facilitates the adaptation of space-grade technology to local contexts, ensuring practical and scalable solutions. By connecting space innovation with terrestrial energy challenges, international cooperation advances the global transition to clean, renewable energy. Space research becomes a cornerstone for sustainable energy strategies that benefit communities and ecosystems worldwide.

Several real-world case studies illustrate the impact of space-derived solar technology on energy efficiency and sustainability. In Germany, advanced solar farms incorporate high-efficiency panels inspired by spacecraft photovoltaic cells. These systems use satellite-based irradiance data to optimize panel orientation and maximize energy production. Tracking technology allows farms to follow the sun throughout the day, increasing electricity generation without additional land or panels. Energy storage solutions adapted from space missions ensure that power is available even during periods of low sunlight. This combination of technology and data demonstrates how lessons from space can optimize renewable energy on Earth.

Urban solar installations have also benefited from innovations originally developed for spacecraft. Cities like Tokyo and Los Angeles have adopted lightweight, flexible solar panels that can be integrated into rooftops, building facades, and public infrastructure. These panels are durable, efficient, and capable of withstanding extreme weather, drawing directly from materials research in space programs. By combining space-inspired design with satellite data on sunlight exposure and environmental conditions, urban planners maximize energy production while minimizing cost and maintenance. These applications provide clean energy in densely populated areas without requiring additional land. Urban solar optimization demonstrates the practical value of space technology for everyday energy solutions.

In rural and off-grid communities, portable solar systems inspired by space missions provide reliable electricity where traditional infrastructure is unavailable. Foldable and lightweight panels, developed for spacecraft, are deployed in remote villages, disaster zones, and refugee camps. These systems generate clean energy for lighting, communication, and essential services, improving quality of life and resilience. Satellite data helps optimize placement, energy output, and maintenance schedules for these portable systems. By applying space-tested technologies, communities gain access to reliable, sustainable power without reliance on fossil fuels. Portable solar solutions thus bring the benefits of space research to some of the most challenging environments on Earth.

Industrial applications also leverage space-derived solar innovations for energy-intensive operations. Factories, warehouses, and large-scale production facilities use high-efficiency panels and tracking systems to reduce electricity costs and carbon emissions. By integrating predictive solar modeling, these facilities can balance energy demand with production schedules and storage capacity. Advanced materials and modular designs allow for scalable installations that meet specific industrial needs. Satellite monitoring ensures optimal system performance and helps identify opportunities for further energy optimization. These applications highlight the broad potential of space research to transform energy usage across sectors.

Case studies in developing countries demonstrate the transformative effect of satellite-informed solar energy deployment. In regions with limited electricity access, space-derived technologies have enabled the creation of microgrids and community energy solutions. Satellites provide data on sunlight availability, environmental conditions, and optimal installation sites, ensuring efficient energy generation. Communities gain reliable electricity for schools, clinics, and small businesses, fostering economic growth and improving living standards. By translating space technology into accessible solar solutions, these projects illustrate the power of innovation to address global energy challenges. Lessons learned from these case studies guide future deployment strategies worldwide.

Space research has proven to be a powerful catalyst for advancing solar energy on Earth, improving efficiency, sustainability, and accessibility. High-efficiency panels, tracking systems, and advanced materials originally designed for spacecraft have been adapted to optimize electricity generation in urban, rural, and industrial settings. Satellites provide critical data on sunlight exposure, weather patterns, and environmental conditions, allowing planners to maximize energy capture and integrate solar power effectively into national grids. Energy storage solutions inspired by space missions ensure reliable power availability, even when sunlight is intermittent. By combining technology, data, and predictive modeling, space research directly enhances the performance and scalability of renewable energy systems worldwide.

The benefits of space-derived solar technology extend to climate mitigation and environmental sustainability. By increasing the efficiency of solar panels and expanding their deployment, greenhouse gas emissions from fossil fuels are reduced, contributing to cleaner air and lower carbon footprints. Urban, industrial, and off-grid communities can transition to renewable energy faster, mitigating the impacts of climate change while supporting sustainable development goals. Satellite monitoring ensures that solar systems are optimized for maximum impact, guiding policy, investment, and infrastructure planning. Space-based innovation thus accelerates the global shift toward clean energy while minimizing environmental harm.

Solar energy inspired by space research also supports energy equity and resilience. Portable and flexible solar systems provide electricity to remote, underserved, or disaster-affected areas, improving quality of life and economic opportunities. Satellite data informs strategic placement and operation of solar installations, ensuring that energy is delivered where it is most needed. By reducing reliance on centralized grids and fossil fuels, these solutions enhance resilience against natural disasters, power outages, and climate variability. Space technology, therefore, empowers communities with reliable, sustainable, and independent energy access.

The integration of space research into global energy strategies promotes innovation and knowledge transfer. Collaborations between space agencies, universities, and private companies have accelerated the development of solar technology, energy storage, and monitoring systems. Lessons learned from space missions guide the design, installation, and operation of solar infrastructure worldwide. This collaborative approach ensures that innovations are scalable, adaptable, and accessible, maximizing benefits for both developed and developing nations. By applying space-based knowledge, humanity can address energy challenges more effectively and equitably.

Ultimately, optimizing solar energy through space research illustrates the transformative power of science and technology in solving global problems. From spacecraft innovations to satellite monitoring and predictive modeling, space-based insights enhance efficiency, sustainability, and resilience across all sectors. Solar energy deployment becomes more effective, cost-efficient, and environmentally responsible, supporting the transition to a low-carbon future. By translating orbital innovation into Earth-bound solutions, space research not only powers our lives but also protects the planet for generations to come. The knowledge, technology, and foresight gained from space continue to illuminate the path toward a sustainable and equitable energy future.

Chapter . 12

Harnessing Space Technology for Precision Agriculture

Applying satellite data to increase crop yields and reduce environmental impact.

Precision agriculture is transforming modern farming by using data-driven techniques to optimize crop production and reduce resource use. Space technology, particularly satellite observations, plays a central role in this transformation. Satellites provide high-resolution imagery, soil moisture data, weather predictions, and crop health monitoring, enabling farmers to make informed decisions. By leveraging these insights, agricultural practices become more efficient, sustainable, and resilient to climate variability. Precision agriculture reduces waste of water, fertilizers, and pesticides while maximizing yield and profit. Space technology ensures that farming adapts to the demands of a growing global population while protecting ecosystems.

Satellite imagery allows farmers to monitor crop health at a detailed level. Sensors detect variations in plant growth, chlorophyll content, and stress caused by pests, diseases, or nutrient deficiencies. Early detection of these issues enables timely interventions, preventing widespread crop loss. Remote sensing also helps identify areas that require targeted fertilization or irrigation, minimizing environmental impact. By continuously monitoring fields, farmers can track the effectiveness of interventions and adjust management strategies accordingly.

Satellite-derived crop health information ensures that agricultural practices are precise, efficient, and environmentally responsible.

Soil moisture and irrigation management benefit directly from space-based observations. Satellites measure soil water content across large areas, providing real-time data for irrigation scheduling. This allows farmers to water crops only when necessary, conserving freshwater and reducing runoff that can harm surrounding ecosystems. Combining soil moisture data with weather forecasts enables predictive irrigation, anticipating dry periods and preventing crop stress. Efficient water management increases yields, lowers operational costs, and reduces pressure on local water resources. Space technology thus ensures sustainable water use while maintaining high agricultural productivity.

Weather prediction and climate monitoring are critical applications of satellite technology in agriculture. Satellites track rainfall, temperature, humidity, and extreme weather events, providing accurate forecasts for farmers. This information supports planting schedules, harvest planning, and risk management strategies. By anticipating adverse weather, farmers can protect crops from frost, drought, or flooding. Integrating satellite weather data with precision farming techniques ensures that resources are used optimally, reducing losses and environmental impact. Space-based weather monitoring allows agriculture to adapt dynamically to changing climate conditions.

Satellite-guided nutrient management enhances both productivity and sustainability. Remote sensing identifies nutrient deficiencies, enabling targeted fertilizer application rather than blanket treatment of entire fields. This precision reduces excess fertilizer runoff, which can pollute waterways and harm biodiversity. Efficient nutrient management also improves soil health over time, supporting long-term agricultural sustainability. By combining satellite data with advanced analytics, farmers can optimize crop growth while minimizing environmental degradation. Space technology provides a foundation for intelligent, sustainable, and highly productive agricultural systems.

Satellite technology significantly improves the monitoring and management of pests and crop diseases. Remote sensing allows farmers to detect early signs of infestations or infections by observing changes in vegetation health and color patterns. Early detection enables targeted interventions, reducing the need for widespread pesticide use and minimizing environmental harm. By tracking the spread of pests and diseases over time, satellite data helps predict outbreaks and inform preventive measures. This proactive approach not only protects crop yields but also reduces costs and supports sustainable farming practices. Space technology ensures that agriculture becomes more precise, efficient, and environmentally responsible.

Yield prediction is another critical application of space-based data in agriculture. Satellites capture information on plant growth, soil conditions, and weather patterns, allowing farmers to estimate expected crop production accurately. Accurate yield forecasts enable better planning for storage, distribution, and market supply, reducing food waste. Governments and organizations can also use satellite-derived data to assess national and regional food security. Predictive models based on satellite observations guide interventions, such as supplemental irrigation or fertilization, to optimize outcomes. By improving yield predictions, space technology helps secure both economic stability for farmers and food availability for communities.

Adaptive farming strategies rely on continuous satellite monitoring to respond to changing conditions. Variations in soil moisture, crop stress, and weather can be tracked in near real-time, allowing farmers to adjust planting schedules, irrigation, and fertilizer application. This flexibility increases resilience against climate variability, extreme weather events, and resource limitations. Satellite data supports decision-making at both local and regional scales, integrating farm-level observations with broader environmental trends. Adaptive strategies minimize waste, protect ecosystems, and maximize productivity. By enabling dynamic and responsive farming, space technology ensures sustainable and resilient agriculture.

Satellite imagery also supports precision planting and field management. High-resolution data identifies soil variability, nutrient levels, and microclimates within fields, guiding the placement of seeds and irrigation infrastructure. Precision planting improves germination rates, reduces resource waste, and optimizes space utilization. Combined with automated machinery and GPS technology, satellites enable highly efficient field operations that save time, labor, and inputs. Farmers can achieve higher productivity with less environmental impact by leveraging space-derived insights for precision planting. Space technology thus bridges the gap between scientific knowledge and practical, scalable agricultural practices.

Data integration from multiple satellite sources enhances decision-making in precision agriculture. Combining soil moisture, crop health, weather forecasts, and topographical data creates a comprehensive understanding of field conditions. Analytical tools and machine learning algorithms process this data, providing actionable recommendations to farmers. Integrated satellite datasets support long-term planning, resource allocation, and sustainability metrics, ensuring that agriculture is productive and environmentally responsible. By unifying diverse observations into a coherent management system, space technology empowers farmers to make informed and strategic decisions. Precision agriculture becomes not only data-driven but also highly adaptive and sustainable.

Real-world case studies demonstrate the effectiveness of satellite-assisted precision agriculture in boosting productivity and sustainability. In the United States, large-scale farms in California's Central Valley use satellite imagery to monitor crop health, soil moisture, and nutrient levels across thousands of acres. By integrating this data with automated irrigation and fertilization systems, farmers reduce water use, minimize chemical runoff, and optimize yields. Early detection of pest infestations and disease outbreaks prevents large-scale crop losses. These practices not only improve economic outcomes but also reduce the environmental footprint of industrial agriculture. Space technology has enabled these farms to achieve higher efficiency while promoting sustainable practices.

Smallholder farms in sub-Saharan Africa also benefit from space-based precision agriculture. Satellites provide data on rainfall patterns, soil fertility, and crop stress, allowing farmers to make informed decisions despite limited access to traditional agricultural resources. Targeted interventions, such as precise irrigation or nutrient application, improve crop yields and food security. Governments and NGOs use satellite data to advise farmers, plan regional agricultural support, and manage scarce water resources. By applying space technology at the local level, small-scale farmers can increase resilience to climate variability and achieve greater productivity. Precision agriculture thus empowers communities with data-driven tools that enhance livelihoods and sustainability.

In India, satellite-assisted farming has been implemented across major rice and wheat-growing regions. Farmers use satellite imagery to assess crop health, track growth stages, and predict harvest outcomes. The data guides irrigation, fertilization, and pest management decisions, reducing waste and environmental impacts. Local agricultural agencies combine satellite data with weather forecasts and soil analysis to provide actionable guidance to farmers. This integrated approach improves yields, supports regional food security, and minimizes ecological stress. Space technology ensures that agricultural systems in India are both productive and adaptive to changing environmental conditions.

Latin America provides another example of satellite-assisted agriculture improving resource efficiency. In Brazil and Argentina, large soybean and maize farms use satellites to monitor soil moisture, detect nutrient deficiencies, and manage irrigation. Satellite data allows farmers to adjust planting patterns, optimize inputs, and respond quickly to climatic variations. This approach reduces fertilizer and water consumption while maximizing output. Environmental benefits include reduced runoff, soil preservation, and protection of surrounding ecosystems. Space technology enables precision agriculture to scale effectively, balancing economic gains with ecological stewardship.

International initiatives leverage satellite technology to support global food security. Organizations such as the Food and Agriculture Organization (FAO) and the Group on Earth Observations (GEO) use satellite data to monitor crop conditions, forecast yields, and identify regions at risk of food shortages. This information informs humanitarian aid, agricultural investment, and policy decisions worldwide. By combining local, regional, and global satellite observations, governments and NGOs can respond proactively to emerging challenges. Space-based precision agriculture thus contributes not only to individual farm productivity but also to the stability and resilience of global food systems.

Integrating satellite-assisted precision agriculture with sustainable farming practices enhances both productivity and environmental stewardship. By using remote sensing to monitor soil health, water use, and crop conditions, farmers can adopt practices that reduce resource waste and maintain ecosystem balance. Precision interventions minimize chemical runoff, prevent soil degradation, and optimize water application, supporting long-term agricultural sustainability. Satellite data also guides crop rotation and cover cropping strategies, improving soil fertility and reducing pest pressures naturally. By combining high-tech monitoring with sustainable methods, agriculture becomes both efficient and ecologically responsible. Space technology thus supports a farming approach that balances yield with environmental preservation.

Climate adaptation is another key benefit of satellite precision agriculture. Farmers face increasing challenges from changing rainfall patterns, temperature fluctuations, and extreme weather events. Satellite observations provide real-time data and predictive models that enable proactive adjustments to planting schedules, irrigation, and pest management. These adaptations reduce crop losses and maintain productivity under adverse conditions. By integrating climate information with field-level monitoring, farmers can anticipate risks and optimize resource allocation.

Space-based insights help agriculture remain resilient in the face of climate variability, securing food production for vulnerable communities.

Water management in precision agriculture is optimized through satellite monitoring. Satellites provide detailed soil moisture data, enabling farmers to apply water only where and when it is needed. This reduces waste, conserves freshwater resources, and maintains healthy soil conditions. Irrigation efficiency improves, particularly in regions with limited water availability or variable rainfall. Combining satellite data with automated irrigation systems ensures precise application that benefits both crops and the environment. Water-smart farming, guided by space technology, strengthens resilience to drought while enhancing productivity.

Biodiversity and ecosystem conservation are supported by satellite-assisted farming. Remote sensing identifies sensitive habitats, buffer zones, and areas at risk from overuse or deforestation. Farmers can avoid damaging critical ecosystems and implement conservation practices such as agroforestry or integrated pest management. By mapping and monitoring land use, satellites help maintain ecological corridors and protect wildlife while sustaining agricultural output. Precision agriculture becomes a tool not only for human benefit but also for environmental stewardship. Space technology ensures that productive land use aligns with ecological preservation goals.

Data-driven decision-making in precision agriculture is empowered by satellites and advanced analytics. Integrating satellite imagery, soil data, weather forecasts, and crop models provides farmers with actionable insights. Analytical platforms translate complex information into recommendations for planting, irrigation, fertilization, and pest control. This holistic approach ensures that resources are allocated efficiently and sustainably. By leveraging space technology and intelligent data systems, agriculture becomes more resilient, productive, and environmentally responsible. Precision farming guided by satellites exemplifies the future of sustainable, adaptive, and technology-driven agriculture.

Satellite-assisted precision agriculture offers transformative benefits for both productivity and environmental sustainability. By monitoring crop health, soil moisture, weather conditions, and nutrient levels, satellites provide farmers with precise, actionable information. This enables targeted interventions that increase yields, reduce water and chemical usage, and minimize environmental impact. Early detection of pests and diseases prevents widespread crop losses, while predictive models improve planting and harvesting decisions. Space technology ensures that agriculture is more efficient, resilient, and capable of meeting the needs of a growing global population. Precision agriculture represents a paradigm shift in farming, combining science, technology, and sustainability.

Global food security is strengthened through satellite-informed farming practices. Accurate yield predictions, adaptive strategies, and resource optimization help stabilize food production in regions facing climate variability, drought, or limited resources. Governments, NGOs, and agricultural organizations can use satellite data to plan interventions, distribute resources, and prevent food shortages. By integrating local, regional, and global observations, satellite technology supports proactive food security strategies. Space-based insights ensure that farmers, communities, and policymakers can respond effectively to emerging challenges. This creates a resilient and sustainable agricultural system that benefits people worldwide.

Water and resource management are enhanced through precision agriculture guided by satellites. Soil moisture data, efficient irrigation, and targeted fertilization reduce overuse and conserve critical resources. Satellite monitoring allows farmers to optimize input use for maximum yield while protecting surrounding ecosystems. By minimizing runoff and soil degradation, space-assisted farming supports long-term environmental health. Water-smart agriculture also improves resilience to drought and climate change, ensuring reliable food production in the face of global pressures. Space technology integrates resource management with high-efficiency farming practices.

Climate adaptation is another crucial advantage of satellite-enabled precision agriculture. Real-time monitoring and predictive models allow farmers to respond proactively to temperature shifts, rainfall variability, and extreme weather events. Adjustments to irrigation, planting schedules, and crop selection help mitigate risks and maintain productivity. Satellite data supports strategic planning that balances short-term yields with long-term sustainability. By aligning farming practices with climate realities, space technology fosters resilient agricultural systems capable of withstanding environmental challenges. Precision agriculture thus empowers farmers to manage uncertainty and secure their livelihoods.

Ultimately, harnessing space technology for precision agriculture demonstrates the power of data-driven, sustainable, and adaptive farming. Satellites provide critical insights that improve productivity, optimize resource use, and protect ecosystems. Farmers can respond to pests, diseases, and climate variability with precision and foresight, while policymakers gain information to strengthen global food security. By integrating observation, prediction, and technology, space-based agriculture ensures that farming remains sustainable, resilient, and efficient. The application of satellite technology to agriculture exemplifies how innovation from space can solve critical Earth-based challenges and support a sustainable future for all.

Chapter . 13

Predicting and Managing Crop Diseases with Remote Sensing

Using orbital data to detect early signs of crop stress and disease to prevent famine.

Remote sensing from satellites has become an essential tool in predicting and managing crop diseases worldwide. Early detection of plant stress, nutrient deficiencies, and disease outbreaks allows farmers to take timely action and prevent widespread crop loss. Satellites capture high-resolution imagery that reveals subtle changes in leaf color, growth patterns, and canopy health. By analyzing these indicators, scientists and farmers can identify potential infections before they escalate. Remote sensing technology reduces the reliance on blanket pesticide applications, saving costs and minimizing environmental impact. Space-based monitoring ensures that agriculture remains productive, sustainable, and resilient against threats to food security.

Monitoring crop health using satellite imagery involves tracking plant vigor, chlorophyll content, and moisture levels. Changes in these parameters can indicate stress caused by pathogens, pests, or environmental factors. By continuously analyzing data from multiple spectral bands, satellites detect anomalies that are invisible to the naked eye. This early-warning capability enables targeted interventions such as localized treatment or improved irrigation. Accurate detection reduces unnecessary pesticide use, protecting ecosystems and human health. Remote sensing thus transforms crop disease management from reactive to proactive, saving both resources and yields.

Weather and environmental data collected by satellites are critical in predicting disease outbreaks. Factors like temperature, humidity, and rainfall influence the proliferation of fungal, bacterial, and viral plant diseases. Satellites provide detailed, real-time environmental data that can be integrated into predictive models for disease risk. By combining plant health indicators with climate conditions, farmers and agricultural agencies can forecast high-risk periods and locations. Early-warning systems allow for preventive measures, including adjusting planting schedules, selecting resistant varieties, or modifying irrigation practices. Space technology enhances the accuracy and timeliness of disease management strategies, reducing the impact on crops and livelihoods.

Satellite-based remote sensing supports precision agriculture by enabling targeted monitoring and intervention. Instead of applying treatments across entire fields, farmers can focus resources on affected areas identified by satellite data. This approach minimizes chemical use, reduces costs, and limits environmental damage. Precision disease management also helps maintain healthy soil and preserves beneficial organisms that support crop growth. By combining remote sensing with ground-based observations and machine learning models, farmers can create highly effective disease management plans. Space technology thus provides the intelligence needed to protect crops while promoting sustainable farming practices.

Global food security benefits directly from the integration of remote sensing into crop disease management. By monitoring large agricultural regions, satellites provide governments, NGOs, and international agencies with early-warning information for potential famine threats. This data allows for preemptive interventions, such as distributing disease-resistant seeds, coordinating treatment campaigns, or adjusting trade and supply policies. Satellite monitoring also supports research on disease patterns, helping scientists develop better prevention strategies and adaptive crop varieties. By detecting problems early and enabling coordinated action, space-based remote sensing reduces the risk of widespread crop failure and hunger.

Satellite-based remote sensing has proven highly effective in detecting and managing crop diseases in various regions around the world. In the United States, precision agriculture programs use satellites to monitor wheat, corn, and soybean fields for signs of fungal infections such as rust and blight. By analyzing multispectral imagery, farmers identify stress patterns before visible symptoms appear. Early intervention with targeted fungicide application or irrigation adjustments prevents crop losses and minimizes chemical use. These programs save millions of dollars annually while protecting ecosystems from excessive pesticide runoff. Space technology thus provides a scalable and proactive solution for disease management in large-scale agriculture.

In sub-Saharan Africa, satellites support smallholder farmers in preventing disease outbreaks that threaten staple crops like maize, sorghum, and cassava. Satellite data combined with weather observations predicts areas at high risk of fungal or viral infections. Governments and NGOs use this information to distribute resistant seed varieties, provide extension services, and coordinate pest management programs. Early-warning systems reduce the impact of diseases on food security and help farmers maintain consistent harvests. By bringing space-derived insights to resource-limited regions, remote sensing strengthens agricultural resilience and community livelihoods. Precision disease monitoring ensures that even small-scale farms can access data-driven protection strategies.

In India, satellite monitoring is applied to rice and wheat cultivation to forecast bacterial and fungal disease outbreaks. High-resolution imagery detects early signs of leaf discoloration, wilting, or canopy thinning, which may indicate infection. Predictive models integrate satellite data with local weather conditions to anticipate the spread of disease. Farmers receive timely alerts that guide treatment decisions, such as targeted pesticide application or adjustments to irrigation. These interventions prevent yield loss and reduce environmental contamination. Space-based monitoring ensures that agriculture in densely populated regions remains productive and sustainable despite disease threats.

South American countries, including Brazil and Argentina, utilize satellite observations to manage diseases in soybean and maize crops. Remote sensing identifies regions experiencing stress due to fungal or viral infections, enabling targeted treatment and preventive measures. This approach reduces chemical usage, protects biodiversity, and improves yield consistency. By integrating satellite data with local agricultural knowledge, farmers can anticipate disease risks and respond effectively. National agricultural programs rely on this data to optimize regional food production and maintain market stability. Space technology enhances disease detection, ensuring healthy crops and sustainable farming practices.

Global initiatives leverage satellite data to monitor crop disease patterns at large scales, supporting food security and policy planning. Organizations such as the Food and Agriculture Organization (FAO) and the Group on Earth Observations (GEO) aggregate satellite-derived information to identify potential threats to key crops worldwide. Predictive models highlight regions at risk, allowing governments and aid organizations to intervene proactively. This large-scale monitoring helps prevent famine, reduce economic losses, and guide research on disease-resistant crops. Remote sensing thus extends the benefits of space technology from individual farms to global food security management.

Integrating satellite data with artificial intelligence (AI) and machine learning has revolutionized crop disease prediction and management. AI algorithms process vast amounts of satellite imagery to detect early signs of stress that may indicate infection. Machine learning models learn from historical data on crop diseases, weather patterns, and soil conditions to predict where and when outbreaks are likely to occur. This combination allows for highly accurate, proactive interventions that prevent crop losses.

Farmers can receive real-time alerts and recommendations, enabling precise treatment and resource allocation. By combining space-based observations with intelligent analysis, agriculture becomes more resilient and efficient.

Predictive analytics allows farmers to anticipate disease spread before it becomes visible on the field. Satellites provide continuous monitoring of crop health indicators, while predictive models simulate the progression of infections based on environmental conditions. These forecasts inform decisions on irrigation, pesticide application, and crop rotation, reducing damage and resource waste. Predictive systems also support regional and national agricultural planning by estimating potential yield losses due to disease. Governments and NGOs use this information to coordinate interventions, distribute resources, and safeguard food security. Space technology, coupled with AI, transforms disease management into a proactive and strategic practice.

Remote sensing combined with drone technology enhances localized disease monitoring. Satellites provide large-scale overviews of crop conditions, while drones capture detailed, high-resolution images for targeted areas of concern. This dual approach allows for precise identification of early-stage infections and efficient treatment applications. Ground-based sensors complement satellite and drone data, creating a multi-layered system of disease detection. Farmers gain actionable insights that improve decision-making and minimize environmental impact. Space-derived technology enables an integrated monitoring network that bridges global observation with local precision.

Machine learning also enables disease classification and risk assessment at scale. By analyzing satellite imagery, models can distinguish between different types of crop stress caused by pests, nutrient deficiencies, or pathogens. This differentiation ensures that interventions are tailored and effective, reducing unnecessary chemical use. Risk maps generated from satellite and model data highlight vulnerable regions, guiding preventive measures and resource allocation. By applying AI to satellite data, farmers and policymakers can prioritize high-risk areas and respond swiftly to emerging threats. Space technology thus supports intelligent, evidence-based agricultural management.

Global collaborations in AI-driven disease monitoring amplify the benefits of satellite remote sensing. Researchers, agricultural organizations, and governments share satellite datasets and predictive models to improve disease detection and response worldwide. International platforms integrate data from multiple satellites, providing comprehensive coverage of major crops and regions. These collaborations enhance early-warning systems, improve intervention strategies, and support global food security efforts. By combining space technology with advanced analytics and international cooperation, humanity gains a powerful tool to combat crop diseases and protect the global food supply.

Case studies from around the world demonstrate the effectiveness of AI-integrated satellite monitoring in managing crop diseases. In Europe, vineyards in France and Italy use satellite imagery combined with predictive models to detect fungal infections such as powdery mildew. Early detection allows targeted spraying, reducing chemical use and protecting the surrounding ecosystem. Data-driven forecasts also help vintners plan irrigation schedules and canopy management to prevent further spread of pathogens. By integrating satellite and AI technologies, European vineyards maintain high-quality yields while minimizing environmental impact. This approach demonstrates the potential for precision disease management in high-value crops.

In Africa, satellite and AI technology supports smallholder farmers in managing maize and sorghum diseases. Satellite imagery provides broad coverage, identifying regions showing stress signals. Machine learning models then predict areas at risk for specific pathogens, such as maize lethal necrosis or rust. NGOs and government agencies deliver targeted interventions, including disease-resistant seeds, localized fungicide application, and community training programs. This approach improves crop resilience, reduces food insecurity, and limits unnecessary chemical use. Space technology, combined with AI, empowers farmers even in resource-limited regions to combat disease threats effectively.

India provides another example of AI-enhanced satellite monitoring in rice and wheat cultivation. Remote sensing detects early signs of bacterial leaf blight, blast, and other crop diseases across large agricultural regions. Predictive models use environmental data to forecast disease spread, enabling proactive management strategies. Farmers receive alerts through mobile applications, guiding treatment application and field management. Early intervention prevents yield losses and reduces costs associated with widespread pesticide use. Space technology, combined with AI, ensures that one of the world's most critical food-producing regions remains productive and sustainable.

In South America, soybean and maize farms in Brazil and Argentina implement satellite and AI technologies to manage fungal and viral diseases. High-resolution imagery identifies early infection, while predictive models determine the likelihood of disease spread. Targeted treatment, optimized irrigation, and crop rotation strategies prevent yield loss and reduce environmental impact. National agricultural programs use satellite-informed data to plan regional interventions, allocate resources, and maintain food security. Space-based technology allows farmers to balance productivity with ecological preservation, ensuring sustainable agriculture at scale.

Global initiatives, such as the Group on Earth Observations (GEO) and the FAO, integrate satellite and AI data to monitor crop diseases across continents. These organizations track disease patterns in staple crops like wheat, maize, and rice, providing early warnings and guidance to countries at risk. By combining multiple satellite sources and predictive models, international agencies can coordinate interventions, prevent regional famine, and support resilient food systems. The integration of space technology with AI-driven monitoring transforms disease management from reactive responses into proactive global strategies, protecting both livelihoods and food security worldwide.

Satellite and AI-driven crop disease management offers transformative benefits for global agriculture. By detecting early signs of stress and infection, farmers can intervene before diseases spread, preventing significant crop losses. Precision monitoring reduces the need for blanket pesticide use, lowering costs and minimizing environmental damage. Predictive models enable strategic planning, integrating weather, soil, and crop data to anticipate disease outbreaks. Space-based technology ensures that farmers can make informed decisions, improving yield, sustainability, and resilience. This proactive approach strengthens agriculture, enabling it to meet the demands of a growing global population.

Global food security is enhanced through satellite-assisted disease prediction. Early-warning systems inform governments, NGOs, and international agencies about potential crop threats, allowing timely interventions. Resources such as disease-resistant seeds, targeted treatments, and extension services can be allocated efficiently to mitigate risks. Monitoring at regional and continental scales ensures that localized outbreaks do not escalate into widespread famine. Space technology, combined with AI analytics, provides the critical intelligence needed to safeguard food supply chains. Farmers, policymakers, and communities benefit from a coordinated and data-driven approach to crop protection.

Environmental sustainability is supported through precision disease management. Targeted interventions reduce overuse of chemicals, protecting soil health, water quality, and biodiversity. Remote sensing allows farmers to minimize unnecessary treatments, preserving beneficial organisms and preventing ecosystem degradation. By integrating satellite monitoring with sustainable practices, agriculture becomes more resilient and environmentally responsible. Space-based technologies provide the tools to balance high productivity with ecological stewardship. Disease management is no longer a reactive struggle but a strategic component of sustainable farming systems.

Efficiency and productivity are improved through data integration from satellites, drones, and ground sensors. Machine learning models analyze multi-source data to optimize treatment schedules, irrigation, and nutrient application. Farmers gain actionable insights that allow them to respond to disease threats quickly and accurately. This efficiency reduces resource consumption, increases yields, and ensures that agricultural operations are economically viable. Space technology and AI collectively transform disease management into a precise, scalable, and sustainable practice.

Ultimately, harnessing satellites and AI for crop disease management illustrates the power of space technology to solve critical Earth-based challenges. By providing early detection, predictive insights, and integrated monitoring, these tools safeguard food production, strengthen global food security, and promote environmental sustainability. Farmers, governments, and international organizations can act decisively, reducing losses, conserving resources, and supporting resilient agricultural systems. The integration of space observation and AI ensures that agriculture can meet the dual demands of productivity and sustainability. Remote sensing represents a transformative solution, demonstrating how innovation from space can directly improve life on Earth.

Chapter . 14

Monitoring Mineral and Energy Resources from Space

Using satellite imaging to identify sustainable extraction sites and reduce ecological damage.

Satellites play a critical role in identifying and monitoring mineral and energy resources on Earth. High-resolution imagery allows geologists and energy companies to detect valuable deposits of minerals, oil, and gas without extensive ground surveys. Remote sensing also tracks environmental changes caused by resource extraction, ensuring that operations minimize ecological impact. By combining satellite data with geological analysis, companies can locate high-potential areas while avoiding sensitive habitats. This approach reduces exploration costs, increases operational efficiency, and promotes sustainable practices. Space-based monitoring ensures that natural resources are managed responsibly, balancing economic development with environmental protection.

Remote sensing technology provides detailed mapping of mineral-rich regions. Satellites equipped with multispectral and hyper-spectral sensors detect the unique spectral signatures of minerals, enabling accurate identification. These maps guide exploration, revealing areas that merit further investigation and highlighting regions that are less viable. By identifying mineral deposits from space, companies can prioritize exploration while reducing unnecessary ground disturbance. This technology also allows governments to plan resource management strategies and regulate extraction more effectively. Space-based mineral mapping ensures that resource development is strategic, efficient, and environmentally conscious.

Energy resource monitoring benefits from satellite observation in multiple ways. Satellites track oil and gas infrastructure, including pipelines, refineries, and storage facilities, providing real-time data on operational efficiency and potential hazards. They also monitor renewable energy sites, such as solar farms and wind installations, to assess performance and detect maintenance needs. Remote sensing supports energy planning by providing insights into resource availability, environmental conditions, and regional energy potential. By combining these observations with predictive analytics, governments and companies can optimize energy extraction and deployment. Space technology thus strengthens energy security while promoting sustainability.

Environmental monitoring is integral to responsible resource management. Satellites track deforestation, soil erosion, water contamination, and habitat disruption associated with mining and drilling activities. Real-time data enables companies and regulators to mitigate negative impacts, enforce compliance, and develop strategies for restoration. By observing changes over time, satellites provide a comprehensive understanding of the ecological consequences of resource extraction. This ensures that economic gains do not come at the expense of environmental integrity. Space-based monitoring supports the development of sustainable resource extraction practices that protect both ecosystems and communities.

Global collaboration in satellite resource monitoring enhances transparency and responsible development. International agencies, governments, and private companies share data to promote sustainable mining and energy practices worldwide. Satellite imagery provides a verifiable record of extraction activities, supporting regulatory compliance and informed decision-making. Collaborative efforts also enable the identification of under-explored regions with potential for sustainable development. By leveraging space technology collectively, stakeholders ensure that natural resources are utilized efficiently, responsibly, and equitably.

Monitoring from space thus becomes a cornerstone of sustainable resource management on a global scale.

Satellite technology allows for precise detection of resource deposits, reducing the need for disruptive ground exploration. Multispectral and hyperspectral sensors identify unique signatures of minerals, oil, and gas, revealing potential extraction sites. By combining satellite imagery with geophysical and geological data, companies can pinpoint high-yield areas efficiently. This reduces the environmental footprint of exploration and minimizes unnecessary excavation. Satellite monitoring also provides early warnings of geohazards such as landslides, sinkholes, or flooding that could affect resource operations. Space-based observation ensures that exploration is safer, more accurate, and less harmful to surrounding ecosystems.

Planning extraction activities benefits greatly from space-derived data. Satellites offer topographical and land-use information to guide infrastructure placement, such as mining roads, drilling platforms, and processing facilities. This enables companies to avoid sensitive ecological zones, waterways, and culturally important areas. By optimizing extraction site layout, satellite monitoring reduces environmental disruption and operational costs. Remote sensing also supports regulatory compliance by providing detailed maps of boundaries, extraction zones, and surrounding environments. Space technology ensures that resource development is planned with both efficiency and environmental stewardship in mind.

Monitoring ongoing operations is another key application of satellite technology. High-resolution imagery tracks the progress of mining, drilling, and renewable energy projects, providing data on land alteration, vegetation loss, and water usage. Companies and regulators can detect unauthorized activities or environmental damage in real time. Satellite monitoring also allows for adaptive management, adjusting extraction methods to reduce ecological impact. By maintaining continuous oversight from space, resource operations become more accountable and transparent. Space technology transforms resource management from reactive enforcement to proactive, data-driven stewardship.

Environmental and social impacts are minimized through satellite-informed decision-making. Remote sensing identifies areas at risk of deforestation, soil degradation, and water contamination caused by extraction activities. Companies can implement mitigation measures, such as reforestation, buffer zones, or water treatment systems, guided by satellite data. Communities impacted by resource operations benefit from better monitoring and planning, reducing risks to livelihoods and local ecosystems. Satellite observation also supports corporate social responsibility initiatives by providing verifiable data on environmental performance. Space technology ensures that resource extraction aligns with sustainability and social equity goals.

Satellite monitoring also supports long-term planning for renewable energy infrastructure. Wind farms, solar arrays, and hydroelectric projects rely on space-based data to assess site suitability, environmental impact, and operational efficiency. Satellites provide insights into terrain, vegetation, sunlight, and wind patterns, optimizing energy production while minimizing ecological disruption. By integrating these observations into energy planning, countries can expand renewable resources responsibly. Space technology thus contributes not only to mineral and fossil energy management but also to sustainable energy development. Monitoring from space ensures that energy and resource projects are efficient, sustainable, and socially responsible.

Several case studies illustrate the effectiveness of satellite-assisted resource monitoring across the globe. In Australia, mining companies use satellite imagery to track iron ore and bauxite extraction sites, ensuring minimal environmental disruption. Multispectral sensors detect vegetation loss, soil erosion, and water contamination near operations. This real-time monitoring allows companies to implement corrective measures promptly, protecting ecosystems and local communities. Satellite-based oversight also supports compliance with environmental regulations and reporting standards. By integrating space technology into daily operations, Australia's mining sector achieves both efficiency and sustainability.

In Canada, oil sands projects rely on satellite monitoring to track land use changes and emissions. High-resolution imagery allows companies to observe extraction sites, tailings ponds, and surrounding ecosystems. Environmental agencies use satellite data to assess potential risks, verify compliance, and enforce remediation measures. By identifying problems early, both companies and regulators can reduce negative impacts on forests, wetlands, and waterways. Space technology enhances accountability and transparency, ensuring that energy development is conducted responsibly. Remote monitoring also supports planning for site restoration after extraction is complete.

Brazil provides another example, where satellite observation monitors both mineral extraction and deforestation in the Amazon region. Satellites detect illegal mining operations, track changes in forest cover, and measure environmental degradation. This data informs law enforcement, government policy, and conservation initiatives. Legal mining operations use satellite insights to optimize extraction methods while minimizing environmental impact. Satellite monitoring ensures that economic development can occur alongside environmental protection. Space technology enables the balance of natural resource exploitation with conservation in one of the world's most sensitive ecosystems.

Renewable energy projects also benefit from satellite-assisted monitoring. In Spain, solar farms and wind energy installations are optimized using space-based data on sunlight exposure, topography, and wind patterns. Satellites track infrastructure development, detect performance issues, and identify potential environmental impacts. By using satellite insights, renewable energy providers improve efficiency, reduce maintenance costs, and minimize disruption to local ecosystems. Space technology ensures that renewable energy development is both sustainable and scalable. The integration of satellite monitoring allows energy projects to meet growing demand while preserving environmental integrity.

In Africa, satellite monitoring supports sustainable resource extraction and energy planning. Countries like South Africa and Namibia use satellites to manage mining operations, monitor environmental compliance, and guide renewable energy development. Remote sensing allows governments to assess resource availability, optimize extraction, and mitigate ecological and social impacts. Satellite data also supports regional energy planning, enabling integration of solar and wind resources into national grids. By leveraging space technology, African nations can pursue economic growth while maintaining environmental sustainability. Global case studies demonstrate that satellites are essential tools for responsible resource management.

Integrating satellite data with artificial intelligence (AI) and geographic information systems (GIS) enhances resource monitoring and extraction efficiency. AI algorithms process satellite imagery to identify patterns in mineral distribution, vegetation changes, and land use. GIS platforms combine these data layers with topography, infrastructure, and environmental constraints, enabling precise planning for extraction sites. By predicting resource locations and potential environmental impacts, companies can optimize operations and reduce ecological disruption. This integration ensures that resource development is both profitable and sustainable. Space technology, combined with advanced analytics, transforms natural resource management into a highly informed and proactive practice.

Predictive analytics helps anticipate environmental and operational challenges before they escalate. Satellites provide continuous monitoring of extraction sites, while AI models analyze trends in soil erosion, water contamination, and habitat loss. Companies and regulators can take preventive measures, such as adjusting mining methods or installing protective infrastructure. Predictive monitoring also minimizes social impacts by identifying communities at risk and guiding interventions. By anticipating challenges, resource managers reduce both economic losses and ecological damage. Space technology thus provides the data-driven foresight necessary for sustainable resource development.

GIS-based analysis of satellite data supports decision-making across multiple sectors. For example, combining satellite imagery with land-use maps helps identify areas suitable for renewable energy installations while avoiding sensitive ecological zones. In mining and oil operations, GIS guides infrastructure planning, waste disposal, and restoration strategies. These integrated systems improve operational efficiency, reduce environmental harm, and ensure regulatory compliance. Satellite-driven GIS platforms allow stakeholders to visualize resource distribution, assess impacts, and plan sustainably. Space technology enables informed decision-making at both local and regional scales.

Environmental compliance benefits from satellite and AI integration. Continuous observation allows governments and companies to verify that extraction activities adhere to environmental regulations. Satellites detect unauthorized land clearing, water contamination, or deforestation in real time. AI algorithms highlight anomalies and prioritize areas requiring inspection or intervention. By maintaining transparency and accountability, these technologies strengthen governance and sustainable management of natural resources. Space-based monitoring ensures that resource development aligns with ecological, social, and economic objectives.

Social and economic planning is enhanced through predictive satellite monitoring. By understanding the distribution of resources, environmental risks, and land-use patterns, governments and companies can plan extraction and energy projects that minimize displacement or disruption of communities. Satellite and AI-driven insights support fair compensation, relocation planning, and infrastructure development. This holistic approach ensures that resource development benefits both industry and society while protecting vulnerable populations. Space technology integrates science, sustainability, and social responsibility to optimize resource management worldwide.

Satellite monitoring of mineral and energy resources provides significant advantages for sustainability, efficiency, and responsible development. By observing land, vegetation, and infrastructure from space, companies and governments can make informed decisions about extraction and energy projects. High-resolution imagery identifies resource-rich areas, monitors environmental impacts, and supports regulatory compliance. This reduces unnecessary ground disturbance, prevents ecological damage, and minimizes social disruption. Space-based monitoring ensures that economic development aligns with environmental stewardship and community well-being. The integration of satellite data into resource management transforms traditional approaches into sustainable, efficient, and accountable practices.

Remote sensing allows for continuous assessment of environmental impacts during extraction operations. Satellites track deforestation, soil erosion, water contamination, and habitat disruption in real time. Early detection of environmental risks enables immediate corrective actions, reducing long-term damage. Companies can optimize extraction methods to minimize their footprint while ensuring productivity. Regulators benefit from transparent, verifiable data that supports enforcement and oversight. Space technology ensures that resource development maintains ecological balance while meeting economic objectives.

Energy resource management also benefits from satellite monitoring. Observations of oil, gas, and renewable energy infrastructure allow for efficient operation, maintenance, and performance optimization. Satellites provide insights into solar exposure, wind patterns, and terrain conditions for renewable energy projects. This data enables strategic planning, efficient resource utilization, and minimal environmental disruption. By combining satellite monitoring with predictive analytics, energy providers can ensure reliable, sustainable, and cost-effective production. Space-based technology is essential for integrating energy development with environmental responsibility.

Social and economic outcomes improve when satellite monitoring guides resource management. Communities near extraction sites benefit from reduced environmental impacts, better planning, and equitable resource distribution. Satellite insights support infrastructure placement, hazard mitigation, and fair compensation policies. Predictive monitoring also helps governments and organizations prioritize interventions to safeguard vulnerable populations. Space technology bridges the gap between economic development and social responsibility, ensuring that resource projects provide broad societal benefits. Remote sensing supports sustainable growth while protecting both people and ecosystems.

Ultimately, monitoring mineral and energy resources from space exemplifies how technology can balance human needs with environmental and social responsibility. Satellites provide detailed, accurate, and timely data that informs decision-making across industries and governments. By combining observation, predictive analytics, and GIS, resource management becomes efficient, sustainable, and transparent. Space-based monitoring safeguards ecosystems, enhances energy security, and promotes global food and resource sustainability. This integration of space technology ensures that the extraction and use of natural resources benefit humanity while protecting the planet for future generations.

Chapter . 15

Space Technology Solutions for Urban Energy Efficiency

Applying lessons from spacecraft energy systems to reduce power consumption in cities.

Urban energy efficiency is becoming increasingly critical as cities worldwide experience population growth and rising energy demand. Space technology, particularly energy systems developed for spacecraft, offers innovative solutions to optimize urban energy use. Satellites provide real-time data on electricity consumption, grid performance, and environmental conditions, enabling cities to manage energy more effectively. Advanced materials, insulation techniques, and energy storage innovations originally designed for space missions can be adapted to urban buildings. By applying these lessons, cities reduce power consumption, lower costs, and decrease carbon emissions. Space technology ensures that urban energy systems are smarter, cleaner, and more sustainable.

Satellites monitor energy consumption patterns across metropolitan areas, providing detailed insights into high-demand zones and inefficiencies. This information allows utilities and city planners to identify areas where energy can be conserved or redistributed. Real-time monitoring also supports the integration of renewable energy sources, ensuring that solar, wind, and hydroelectric power are used efficiently. By analyzing urban energy flow from space, municipalities can implement targeted energy-saving measures such as smart lighting, dynamic heating, and cooling adjustments. Space-based monitoring transforms urban energy management from reactive to proactive, maximizing efficiency while minimizing environmental impact.

Energy storage solutions inspired by spacecraft technology are critical for urban efficiency. Space missions rely on compact, high-capacity batteries and thermal management systems to store and regulate energy for spacecraft operations. These innovations can be applied to city grids, enabling efficient storage of solar and wind energy. Distributed energy storage allows for load balancing during peak demand, reducing strain on infrastructure and avoiding power outages. By integrating space-derived storage solutions, urban areas can maintain reliable electricity while reducing reliance on fossil fuels. Space technology thus supports both energy resilience and sustainability in cities.

Building efficiency can also benefit from materials and systems developed for space. Advanced insulation, reflective coatings, and temperature regulation technologies originally designed for spacecraft can be adapted to urban structures. Smart building management systems, informed by satellite data, monitor temperature, humidity, and occupancy to optimize energy use. These adaptations reduce heating, cooling, and lighting demands, lowering overall energy consumption. By combining space-based materials innovation with intelligent monitoring, cities can achieve higher energy efficiency without compromising comfort or functionality. Space technology transforms urban architecture into a model of sustainable energy use.

Urban planning benefits from integrating satellite data with energy management strategies. Remote sensing provides insights into city layouts, infrastructure density, and environmental conditions, allowing planners to optimize energy distribution and transportation networks. By identifying areas of high energy consumption or potential renewable energy integration, satellite data guides infrastructure investment and policy development. This holistic approach ensures that cities are designed for efficiency, sustainability, and resilience. Space technology enables urban planners to implement evidence-based strategies that reduce energy waste and support greener, more livable cities.

Several cities worldwide are applying satellite and space-inspired technologies to improve energy efficiency. In Singapore, high-resolution satellite imagery is used to monitor building energy consumption, assess solar potential, and optimize urban planning. Smart grid systems, informed by space-based data, adjust energy distribution in real time to reduce peak load and prevent blackouts. Advanced materials inspired by spacecraft are integrated into building designs to improve insulation and thermal regulation. This combination of satellite monitoring, smart infrastructure, and space technology reduces overall energy demand while enhancing urban sustainability. By applying lessons from space systems, Singapore demonstrates how data-driven urban energy management can benefit both the environment and citizens.

In Copenhagen, Denmark, satellites are used to monitor urban energy use and environmental impact in real time. Remote sensing helps identify energy-intensive areas, guide efficiency improvements, and optimize district heating systems. Space-based insights are integrated with smart building management and IoT sensors to create responsive energy systems. These technologies allow for precise control of lighting, heating, and cooling, reducing unnecessary consumption. Copenhagen's urban energy strategies also leverage predictive analytics from satellite data to anticipate peak demand periods. Space technology thus enables the city to achieve ambitious energy reduction goals while maintaining comfort and functionality.

New York City uses satellite monitoring to optimize energy efficiency in commercial and residential buildings. Remote sensing identifies rooftop solar potential, monitors energy consumption patterns, and tracks environmental conditions such as heat islands. Space-inspired energy storage solutions and smart grid systems help manage fluctuating electricity demand efficiently. Satellite data also supports urban planning decisions, including zoning and infrastructure investments, to reduce energy waste. By integrating space technology into urban energy strategies, New York increases resilience, lowers emissions, and ensures reliable power distribution. The combination of observation, analytics, and innovation transforms energy management into a proactive process.

Tokyo, Japan, employs satellites and space-derived technologies to manage energy across its dense metropolitan environment. Satellite data helps monitor electricity consumption, evaluate renewable energy integration, and identify infrastructure inefficiencies. Advanced materials developed for spacecraft, such as reflective coatings and insulation, are applied in buildings to reduce heating and cooling needs. Smart grids, powered by real-time data, balance load and optimize energy distribution. These efforts reduce greenhouse gas emissions while maintaining the city's high standard of living. Space technology ensures that energy systems in Tokyo are both efficient and environmentally responsible.

In Melbourne, Australia, satellite imagery supports urban energy planning and sustainable infrastructure development. Remote sensing data identifies energy-intensive areas and evaluates potential renewable energy projects. Smart building management systems, guided by space technology, monitor occupancy, lighting, and climate control to minimize consumption. Satellite data also informs transportation planning, reducing traffic-related energy use and emissions. By leveraging space technology, Melbourne achieves a comprehensive and integrated approach to urban energy efficiency. Cities worldwide can replicate these strategies to create sustainable, data-driven energy systems.

Integrating renewable energy sources into urban grids is greatly enhanced by space-based monitoring. Satellites provide real-time data on sunlight exposure, wind patterns, and environmental conditions, enabling precise placement of solar panels and wind turbines. This ensures maximum energy generation while reducing waste and inefficiencies. Predictive models, powered by satellite data, forecast energy production and consumption, allowing cities to balance supply and demand effectively. By combining renewable energy with space-inspired storage solutions, cities can reduce dependence on fossil fuels. Space technology ensures that urban renewable energy systems are both reliable and environmentally sustainable.

Predictive modeling using satellite data supports energy demand management and load balancing in cities. AI algorithms analyze historical consumption patterns, weather forecasts, and population density to anticipate peak energy periods. This allows utilities to optimize energy distribution and reduce strain on the grid. Satellites also monitor environmental factors, such as urban heat islands, that influence energy demand. Predictive insights enable cities to implement targeted interventions, such as demand-response programs and dynamic pricing. Space-based modeling ensures that urban energy systems operate efficiently, reducing waste and carbon emissions.

Energy-efficient lighting systems benefit from satellite and space technology integration. Satellites identify areas of high nighttime energy consumption, guiding the deployment of smart street lighting. These systems use sensors to adjust brightness based on pedestrian and vehicle traffic, reducing unnecessary electricity use. Space-derived insights also inform the placement of public lighting to minimize light pollution while maintaining safety. By combining real-time monitoring with adaptive technologies, cities can achieve significant energy savings. Space technology ensures that lighting infrastructure is sustainable, cost-effective, and environmentally responsible.

Transportation energy efficiency is enhanced through satellite monitoring and data analysis. Satellites track traffic patterns, congestion hotspots, and vehicle emissions, providing cities with information to optimize public transit, traffic flow, and road design. Predictive analytics helps plan peak-hour routing, reduce idling, and improve fuel efficiency. Satellite-informed urban planning supports the integration of electric vehicle charging infrastructure and shared mobility solutions. By leveraging space technology, cities can reduce transportation-related energy consumption and lower greenhouse gas emissions. This contributes to cleaner, more sustainable urban environments.

Smart building systems integrate satellite monitoring, predictive analytics, and energy management to optimize urban energy use. Satellites provide data on sunlight, temperature, and weather, which inform heating, cooling, and lighting adjustments. AI algorithms analyze occupancy patterns, energy consumption, and environmental conditions to minimize waste. Buildings can adapt in real time, reducing energy demand while maintaining comfort. Space technology-inspired materials, such as advanced insulation and reflective coatings, further enhance efficiency. Together, these innovations create urban spaces that are sustainable, resilient, and energy-conscious.

Several cities worldwide provide exemplary models of integrated urban energy systems using satellite and space-inspired technologies. In Amsterdam, the combination of satellite monitoring and smart grid infrastructure optimizes energy distribution across residential and commercial areas. Satellites track solar energy potential, energy demand, and weather patterns, guiding the allocation of renewable power. AI-driven predictive models help balance load, prevent outages, and reduce reliance on fossil fuels. Advanced building materials inspired by spacecraft technology further enhance energy efficiency in homes and offices. By integrating these tools, Amsterdam achieves a sustainable urban energy system that reduces environmental impact and costs.

In Seoul, South Korea, urban energy efficiency programs leverage satellite data for comprehensive monitoring and planning. Satellites observe energy consumption patterns, identify inefficiencies, and assess renewable energy potential across the city. Smart buildings equipped with adaptive lighting, heating, and cooling systems respond to real-time satellite and sensor data. The city also integrates predictive analytics to anticipate peak energy demand and optimize grid performance. Space-inspired energy storage solutions improve reliability and allow for the integration of intermittent renewable sources. Seoul's approach demonstrates the potential of combining space technology, AI, and smart infrastructure to create efficient, resilient cities.

Barcelona, Spain, uses satellite monitoring and smart grid technology to improve urban energy efficiency. High-resolution satellite imagery identifies high-consumption areas, informs renewable energy integration, and supports building retrofits.

Predictive models guide energy distribution and resource planning, reducing peak demand and overall consumption. Space-derived materials and innovative insulation systems are incorporated into new and existing buildings to reduce heating and cooling needs. Public infrastructure, including street lighting and transportation networks, is optimized using satellite insights to lower energy usage. Barcelona exemplifies how urban energy management can be data-driven, sustainable, and scalable.

In Vancouver, Canada, city planners combine satellite monitoring with green building initiatives to enhance energy efficiency. Satellites provide detailed information on sunlight, temperature, and environmental conditions that inform building design and urban planning. Smart energy systems regulate heating, ventilation, and lighting in real time, reducing waste and costs. Satellite-derived data also supports the deployment of rooftop solar panels and integration of energy storage systems. By linking observation, predictive analytics, and advanced materials, Vancouver creates an urban energy ecosystem that is both sustainable and resilient. Space technology ensures the city meets its environmental goals while maintaining high standards of living.

Singapore applies an integrated approach that combines satellite monitoring, smart grids, and renewable energy planning to maximize urban energy efficiency. Satellite data informs infrastructure development, building retrofits, and renewable energy site selection. Predictive models and AI manage energy distribution in real time, optimizing consumption and reducing waste. Advanced materials and energy storage solutions inspired by spacecraft technology enhance performance and reliability. This comprehensive strategy ensures that Singapore's urban energy systems are efficient, environmentally responsible, and prepared for future growth. Space technology plays a pivotal role in transforming cities into sustainable, high-performance urban centers.

Satellite and space-inspired technologies provide cities with powerful tools to enhance energy efficiency, reduce consumption, and promote sustainability. By monitoring energy use, environmental conditions, and renewable energy potential from space, cities gain actionable insights for smarter urban planning. Advanced materials, energy storage, and adaptive building systems derived from spacecraft innovations further optimize urban energy use. Smart grids and predictive analytics enable real-time load balancing, peak demand management, and integration of renewable energy sources. Space technology transforms urban energy systems into resilient, cost-effective, and environmentally responsible networks. By applying these innovations, cities can meet growing energy demands while minimizing their carbon footprint.

Urban energy efficiency directly contributes to climate resilience. Satellites monitor urban heat islands, weather patterns, and environmental changes that affect energy demand. This data informs strategies to reduce heat-related energy consumption and manage cooling and heating needs efficiently. Renewable energy integration, optimized through space-based insights, reduces reliance on fossil fuels and lowers greenhouse gas emissions. Energy-efficient urban design, informed by satellite and spacecraft technology, improves resilience to climate variability and extreme weather events. Space technology ensures that cities can adapt to climate challenges while maintaining sustainable energy practices.

Economic benefits arise from implementing satellite-guided energy efficiency programs. Reduced energy consumption lowers operational costs for businesses, government institutions, and households. Smart grids and predictive energy management systems optimize resource allocation and prevent costly power outages. Renewable energy integration guided by satellite data reduces dependence on imported fuels and stabilizes energy prices. Advanced materials and adaptive building systems extend the lifespan of infrastructure, reducing maintenance and replacement costs. Space technology enables cities to achieve both economic savings and environmental sustainability simultaneously.

Social and environmental outcomes improve when urban energy systems are designed using satellite and space-based innovations. Efficient energy use reduces pollution and greenhouse gas emissions, improving air quality and public health. Equitable distribution of energy resources ensures that all communities have access to reliable, affordable power. Predictive monitoring and smart infrastructure reduce energy waste, enhancing quality of life and supporting sustainable urban growth. Space technology ensures that urban energy planning prioritizes people, ecosystems, and long-term resilience. Cities become more livable, equitable, and environmentally responsible through the application of space-inspired solutions.

Ultimately, harnessing space technology for urban energy efficiency demonstrates the transformative potential of innovation from orbit applied on Earth. Satellites, smart grids, advanced materials, and predictive analytics collectively enable cities to optimize energy use, integrate renewable resources, and reduce environmental impact. Urban areas become more sustainable, resilient, and prepared to meet future energy demands while mitigating climate change. By learning from spacecraft systems and leveraging space-based observation, cities worldwide can create energy-efficient, environmentally responsible, and technologically advanced urban environments. Space technology ensures that urban energy management is both innovative and sustainable for generations to come.

Part IV: Health, Safety, and Disaster Management

Chapter . 16

Improving Public Health Through Satellite Data

Tracking disease outbreaks globally and responding faster with space-enabled technology.

Public health is increasingly reliant on satellite technology to monitor, predict, and respond to disease outbreaks worldwide. Satellites provide real-time data on environmental conditions, population density, and movement patterns, which influence the spread of infectious diseases. By combining satellite imagery with epidemiological models, health authorities can identify high-risk areas and allocate resources effectively. Remote sensing also helps track vector populations, such as mosquitoes that carry malaria or dengue, by monitoring rainfall, temperature, and vegetation. Space-enabled technologies allow for early warning systems that reduce response times and save lives. Integrating satellite data into public health strategies ensures more accurate, timely, and effective interventions.

Satellite observation enables monitoring of environmental factors that contribute to disease emergence. Temperature, precipitation, humidity, and vegetation patterns can all influence the transmission of diseases. For instance, satellite data identifies regions prone to waterborne or vector-borne illnesses by tracking flooding, stagnant water, and urban heat islands. By mapping these environmental conditions, public health agencies can predict outbreaks and implement preventive measures, such as targeted vaccinations or vector control programs. Remote sensing allows authorities to focus interventions where they are needed most, improving efficiency and reducing unnecessary resource use. Space technology thus strengthens the link between environmental monitoring and disease prevention.

Tracking human mobility and population density is another critical application of satellite data for public health. Satellites monitor urbanization, migration, and travel patterns, which affect how diseases spread between communities and across borders. This information helps health authorities anticipate potential hotspots, prepare hospitals and clinics, and plan vaccination campaigns. Satellite data also supports real-time monitoring during epidemics, enabling authorities to adjust strategies dynamically. By integrating mobility patterns with epidemiological data, health agencies gain a comprehensive view of risk and transmission dynamics. Space-enabled insights improve the accuracy and timeliness of public health decision-making.

Satellite data supports global surveillance systems for infectious diseases. International organizations, such as the World Health Organization (WHO), leverage satellite imagery to monitor environmental conditions that may trigger outbreaks. Data from space assists in tracking diseases such as cholera, Zika, Ebola, and influenza, providing early-warning alerts to countries and regional health networks. These insights allow for proactive responses, including deploying medical teams, distributing supplies, and educating at-risk communities. By combining satellite observation with on-the-ground reporting, global health authorities create an integrated monitoring system. Space technology thus enhances preparedness and response at local, national, and international levels.

Beyond infectious diseases, satellites help monitor air quality and environmental exposures that impact public health. Remote sensing detects pollutants, particulate matter, and greenhouse gas concentrations, providing critical information for urban planning and health interventions. Areas with poor air quality can be identified and addressed through policy measures or targeted health campaigns. Satellite data also supports research on the links between environmental conditions and chronic diseases, such as asthma or cardiovascular illness. By combining environmental monitoring with epidemiological analysis, public health agencies can prevent illness and promote healthier communities. Space technology provides a comprehensive toolset to protect human health through observation, prediction, and intervention.

Several case studies demonstrate the power of satellite data in improving public health outcomes around the world. In Africa, satellite imagery has been used to predict malaria outbreaks by monitoring rainfall, temperature, and vegetation, which influence mosquito breeding. Health authorities can pre-position bed nets, insecticides, and medical supplies in areas most at risk. By combining satellite observations with local epidemiological data, interventions become more targeted and efficient. This proactive approach reduces disease transmission, saves lives, and maximizes the use of limited resources. Space technology enables a shift from reactive response to anticipatory action in controlling infectious diseases.

In South Asia, satellite monitoring has been applied to track the spread of dengue fever and cholera in densely populated urban areas. Satellite data identifies flood-prone zones, areas with stagnant water, and high-risk neighborhoods. Public health teams use this information to deploy vaccination drives, clean water initiatives, and vector control measures. Real-time satellite observation allows authorities to adjust strategies as conditions change, improving effectiveness. By integrating environmental and population data from space, cities can prevent outbreaks and protect vulnerable communities. Space-enabled insights provide a critical tool for disease prevention in highly populated regions.

During the Ebola outbreak in West Africa, satellite data helped track population movements and environmental conditions that influenced disease spread. Remote sensing revealed areas with limited access to healthcare and regions where transmission risk was highest due to human mobility and environmental factors. Health agencies used this information to plan interventions, allocate medical personnel, and establish quarantine zones. By integrating satellite data with field reporting, officials could anticipate new outbreak locations and respond faster. Space technology played a crucial role in supporting coordinated and informed public health responses.

In Latin America, satellite-assisted monitoring has been instrumental in controlling Zika virus transmission. Satellites track rainfall, temperature, and urban water accumulation to map mosquito habitats. Health authorities use this data to plan targeted spraying campaigns, community education programs, and vector control interventions. Predictive models based on satellite observations forecast future risk areas, allowing preemptive action. By combining satellite data with epidemiological surveillance, countries reduce infection rates and mitigate the impact of epidemics. Space technology strengthens public health infrastructure and enables rapid, precise interventions.

Globally, satellite networks support international disease surveillance and response coordination. Organizations such as the World Health Organization (WHO) and the European Centre for Disease Prevention and Control (ECDC) use satellite data to monitor environmental factors that influence disease outbreaks. Satellite-derived insights allow timely alerts to countries at risk, guiding preventive measures and resource allocation. By sharing space-based data, international agencies enhance preparedness, reduce transmission, and improve health outcomes worldwide. Space technology ensures that public health efforts are evidence-based, proactive, and globally coordinated.

Integrating satellite data with artificial intelligence (AI) and machine learning enhances the accuracy and speed of disease outbreak prediction. AI algorithms process vast amounts of satellite imagery alongside epidemiological and environmental data to detect early warning signs of outbreaks. Machine learning models identify patterns, such as correlations between rainfall, temperature, vegetation, and disease prevalence. By predicting where and when diseases are likely to emerge, health authorities can implement preventive measures in advance. This proactive approach reduces the spread of infection and maximizes the effectiveness of public health interventions. Space technology, combined with AI, transforms disease management into a predictive, data-driven process.

Predictive modeling using satellite data allows public health agencies to allocate resources efficiently. Models forecast regions at risk for outbreaks and identify the populations most vulnerable to disease exposure. Hospitals, clinics, and medical supply chains can prepare accordingly, ensuring sufficient personnel, vaccines, and equipment are available. Satellite-derived insights also guide public awareness campaigns, targeting communities before disease spreads widely. By anticipating healthcare needs, governments and NGOs reduce both morbidity and mortality rates. Space-enabled predictive tools turn reactive healthcare responses into proactive strategies that save lives.

AI-enhanced satellite monitoring also supports vector control and environmental health programs. Satellites track mosquito breeding habitats, water contamination, and urban heat islands that influence the prevalence of infectious diseases. Machine learning models analyze these patterns to optimize spraying campaigns, drainage improvements, and sanitation initiatives. Early detection of environmental risks allows for timely interventions that prevent disease transmission. By combining remote sensing with intelligent analytics, public health programs become more precise, cost-effective, and sustainable. Space technology ensures that interventions target the right locations at the right time.

Integration of satellite data into global health networks enhances disease surveillance and outbreak response. International agencies use satellite imagery to monitor cross-border disease risks, track environmental changes, and coordinate interventions. Machine learning models aggregate data from multiple sources to identify emerging threats in real time. This collaboration allows countries to respond swiftly to potential epidemics, preventing localized outbreaks from becoming global crises. Space-based monitoring thus strengthens global health security and ensures coordinated responses to emerging health threats. By leveraging AI and satellite data, public health systems become more resilient and effective.

Beyond infectious disease, satellites assist in monitoring chronic disease risk factors and environmental exposures. Remote sensing tracks air quality, pollution levels, and urban heat islands that influence respiratory, cardiovascular, and metabolic diseases. Machine learning models identify populations at risk and predict health outcomes based on environmental conditions. This enables targeted public health interventions, urban planning improvements, and policy development to reduce chronic disease burden. By integrating satellite data with AI, governments and healthcare providers can proactively address both acute and long-term health challenges. Space technology provides the tools to safeguard human health comprehensively.

Case studies around the world highlight the impact of integrating satellite data with AI in public health interventions. In Kenya, satellite monitoring of rainfall and vegetation combined with machine learning models predicts malaria outbreaks with high accuracy. Health authorities can pre-position medical supplies, deploy mosquito nets, and plan targeted spraying campaigns before infections peak. This proactive strategy reduces disease incidence and optimizes the use of limited healthcare resources. Satellite-based early warning systems empower local communities to take preventive measures. Space technology, paired with intelligent analytics, transforms malaria management from reactive to predictive.

In Brazil, AI-driven satellite monitoring supports dengue fever prevention in urban areas. Satellites detect standing water, temperature variations, and urban heat islands, which influence mosquito breeding. Predictive models forecast high-risk neighborhoods, guiding public health interventions, such as localized vector control and public awareness campaigns. Real-time satellite data allows authorities to adapt strategies as conditions change, minimizing outbreaks. By combining AI and remote sensing, Brazil reduces the spread of dengue while improving resource allocation. Space-enabled insights ensure that public health actions are both timely and precise.

During the Ebola outbreak in West Africa, satellite and AI technologies enhanced epidemic response and containment. Satellites tracked human movement, population density, and environmental factors, helping predict where new cases were likely to emerge. Machine learning models integrated this data with field reports to prioritize resource deployment, medical personnel, and treatment facilities. Real-time monitoring facilitated rapid intervention and reduced disease transmission across borders. By leveraging space technology, health authorities could respond faster and more effectively to a highly contagious virus. AI and satellite integration became a cornerstone of strategic epidemic management.

In India, satellite-based AI systems monitor air quality, heat stress, and water contamination, which impact public health. Predictive models forecast areas at risk for respiratory illnesses, heat-related deaths, and waterborne diseases. Authorities use this information to issue public alerts, plan healthcare resources, and implement preventive measures. Satellite monitoring allows for large-scale, real-time tracking across urban and rural regions, increasing the efficiency of public health responses. This integration of space technology and AI ensures that interventions reach those most vulnerable. Public health systems become proactive and targeted, reducing illness and mortality rates.

Globally, international organizations leverage satellite and AI technologies to coordinate disease surveillance and emergency response. The World Health Organization (WHO) and other agencies integrate satellite-derived data with predictive models to track epidemics and environmental health risks. Early-warning alerts guide vaccination campaigns, vector control programs, and emergency healthcare deployments across continents. By sharing satellite data internationally, countries can anticipate outbreaks and implement coordinated interventions. Space technology, combined with AI, strengthens global health infrastructure and provides the tools necessary to prevent local health crises from escalating into worldwide emergencies.

Satellite and AI-driven public health monitoring transforms disease prevention, healthcare planning, and global health security. By providing real-time environmental and population data, satellites enable early detection of disease risks, guiding timely interventions. Predictive models inform the allocation of medical personnel, supplies, and preventive measures, reducing morbidity and mortality. Integration of satellite data with machine learning allows health authorities to anticipate outbreaks, identify high-risk populations, and optimize resource distribution. Space technology ensures that public health responses are proactive, precise, and data-driven. The combination of observation, prediction, and analytics strengthens public health infrastructure at local, national, and global levels.

The ability to monitor environmental factors that influence health is critical for both infectious and chronic diseases. Satellites track air quality, temperature, rainfall, and water quality, which directly impact human well-being. Predictive models forecast disease risks and environmental hazards, enabling preventive actions before health crises occur. Urban planners, policymakers, and healthcare providers use these insights to design healthier, safer communities. Satellite-based monitoring ensures that environmental exposures are managed effectively to reduce health risks. Space technology provides a comprehensive toolset for promoting population health and sustainability.

Global collaboration is enhanced by satellite-assisted public health systems. International agencies, governments, and NGOs share satellite data and predictive models to improve surveillance and intervention strategies. Coordinated efforts allow for timely response to outbreaks, efficient resource allocation, and containment of epidemics. Satellite monitoring ensures transparency and accuracy, supporting trust and collaboration among nations. By leveraging space technology collectively, the global community can strengthen preparedness and resilience against emerging health threats. Space-enabled insights enhance coordination and effectiveness in protecting human health worldwide.

Economic benefits also arise from satellite-informed public health strategies. Early detection of disease outbreaks reduces healthcare costs by minimizing hospitalizations, treatments, and emergency responses. Preventive interventions guided by satellite data are more efficient and targeted, reducing wasteful spending on ineffective measures. Healthier populations contribute to increased productivity, economic stability, and social well-being. Space technology thus supports both economic and public health goals simultaneously. Investing in satellite monitoring systems strengthens resilience, reduces costs, and enhances overall societal welfare.

Ultimately, improving public health through satellite data demonstrates the profound impact of space technology on life on Earth. By integrating real-time observation, AI, predictive modeling, and global collaboration, health authorities can prevent disease, manage environmental risks, and respond swiftly to emerging threats. Satellite-assisted monitoring ensures that interventions are timely, precise, and effective, saving lives and resources. Space technology provides a critical framework for sustainable, resilient, and equitable public health systems. The integration of these tools represents a transformative solution, showing how innovation from space can safeguard humanity on the ground.

Chapter . 17

Air Quality Monitoring with Space-Based Sensors

Using satellites to measure pollutants and guide policy for cleaner air worldwide.

Air pollution is one of the most pressing public health and environmental challenges of our time, and space-based sensors have become a critical tool in monitoring and managing air quality. Satellites equipped with advanced spectrometers detect concentrations of pollutants such as nitrogen dioxide, carbon monoxide, sulfur dioxide, ozone, and particulate matter. This information provides a comprehensive view of air quality over large regions, including urban centers, industrial zones, and remote areas. By combining satellite observations with ground-based monitoring, scientists can track pollution trends, identify sources, and assess the effectiveness of emission-reduction strategies. Space technology enables accurate, real-time monitoring that supports informed decision-making for public health and environmental management. Air quality monitoring from orbit ensures that interventions are timely, precise, and data-driven.

Satellite observations allow for the identification of pollution hotspots and tracking of pollutant transport across borders. Winds, temperature inversions, and atmospheric circulation can carry pollutants far from their source, affecting regions downwind. Space-based sensors capture these dynamics, helping authorities understand how air quality changes over time and across locations. By mapping pollution sources and dispersion patterns, policymakers can implement targeted regulations and interventions. International collaboration is enhanced as countries gain shared insights into transboundary air pollution. Satellites provide the comprehensive data necessary to manage air quality at local, national, and global scales.

Air quality monitoring from space supports the evaluation of policy measures and emission-reduction strategies. For example, satellite data can track changes in nitrogen dioxide levels before and after the implementation of clean energy regulations or traffic reduction programs. This allows governments to measure the effectiveness of interventions and adjust policies accordingly. By providing an independent and objective assessment, satellite monitoring reinforces accountability and transparency. Public awareness is also improved, as satellite imagery and pollution maps inform communities about air quality trends. Space technology thus ensures that environmental policies are evidence-based and results-oriented.

In addition to traditional pollutants, satellites detect greenhouse gases such as carbon dioxide and methane, which contribute to climate change. Monitoring emissions at regional and global scales informs both climate policy and air quality management. By identifying sources of greenhouse gases, such as industrial facilities, agriculture, or transportation, satellite data helps governments prioritize mitigation efforts. Space-based monitoring allows for long-term trend analysis, showing the impact of interventions over months and years. Integrating air quality and climate data ensures a holistic approach to environmental management and sustainable development. Space technology connects local pollution control to global climate action.

Public health applications of satellite air quality monitoring are significant. Exposure to pollutants is linked to respiratory illnesses, cardiovascular disease, and premature death. By identifying areas with high pollution levels, health authorities can issue alerts, advise vulnerable populations, and plan medical resources. Urban planners can use satellite data to design healthier cities, including green spaces, traffic management, and industrial zoning. Real-time monitoring also supports emergency response to events such as wildfires or industrial accidents that degrade air quality rapidly.

Space technology ensures that decisions affecting human health are informed by accurate, up-to-date environmental data.

Several cities and regions have successfully leveraged satellite monitoring to improve air quality and protect public health. In Beijing, China, satellites track nitrogen dioxide, particulate matter, and other pollutants emitted by industrial activities and vehicles. This data informs government regulations, traffic restrictions, and industrial emission reduction programs. By combining satellite monitoring with ground-based sensors, Beijing can identify pollution hotspots and assess policy effectiveness. Real-time observations enable authorities to issue public alerts during periods of high pollution, protecting vulnerable populations. Space-based monitoring has become a cornerstone of Beijing's air quality management strategy, ensuring data-driven and responsive interventions.

In Los Angeles, USA, satellite data is used to monitor smog and ozone levels across the metropolitan area. Space-based sensors detect patterns of pollutant accumulation influenced by traffic, industrial activity, and meteorological conditions. This information supports urban planning decisions, such as the placement of green infrastructure, traffic flow adjustments, and industrial zoning. Satellite monitoring also informs public advisories during periods of high pollution, reducing exposure risks. By integrating remote sensing with predictive modeling, Los Angeles can anticipate pollution spikes and implement preemptive measures. Space technology allows the city to manage air quality proactively, protecting public health while guiding policy decisions.

Delhi, India, relies on satellite monitoring to track severe air pollution, particularly during winter months when smog becomes hazardous. Satellites detect particulate matter and pollutants originating from vehicles, industry, and crop burning in nearby regions. Government agencies use this information to enforce temporary restrictions on vehicle use, regulate industrial emissions, and manage waste burning. Satellite-based data also supports long-term strategies, such as transitioning to cleaner fuels and promoting public transportation. By providing a comprehensive view of regional air quality, satellites empower authorities to make informed decisions that reduce health risks. Space technology ensures that interventions are timely, targeted, and effective.

In Europe, the European Space Agency's Sentinel satellites provide high-resolution air quality data across multiple countries. These satellites measure nitrogen dioxide, sulfur dioxide, ozone, and particulate matter, enabling cross-border collaboration in pollution management. Countries can identify transboundary pollution sources and develop coordinated emission reduction strategies. Satellite data also supports research into the health impacts of air pollution, guiding public health policy and urban planning. By integrating space-based monitoring into environmental governance, Europe improves air quality, reduces health risks, and promotes sustainable development. Space technology facilitates international cooperation and evidence-based decision-making.

Air quality monitoring from space has also been crucial during wildfire events, such as those in California and Australia. Satellites detect smoke plumes, particulate matter concentrations, and their movement across regions. Public health authorities use this data to issue evacuation orders, distribute protective equipment, and advise on indoor air safety. Real-time satellite monitoring allows rapid response to environmental emergencies, minimizing health impacts on affected populations. By combining satellite data with predictive models, authorities can anticipate air quality deterioration and prepare interventions in advance. Space technology ensures that emergency air quality management is precise, timely, and effective.

Integrating satellite air quality data with artificial intelligence and predictive modeling significantly enhances pollution management strategies. AI algorithms process large volumes of satellite imagery to identify trends, anomalies, and emerging pollution hotspots. These systems can distinguish between pollution sources such as traffic, industry, wildfires, and natural dust events. Predictive models use historical satellite data combined with weather forecasts to anticipate pollution spikes days or even weeks in advance. This foresight allows governments and city planners to implement preventive measures before air quality deteriorates. Space technology transforms air quality management from reactive monitoring into proactive prevention.

Predictive air quality models support targeted policy interventions and efficient resource allocation. When satellites indicate rising pollution levels, authorities can activate temporary traffic restrictions, adjust industrial operations, or promote public transportation. Health agencies can prepare hospitals for increased respiratory cases and issue advisories to vulnerable populations. Satellite-based forecasting also supports long-term urban planning by identifying chronic pollution zones. These insights guide infrastructure investments such as green spaces, low-emission zones, and renewable energy systems. Space-enabled prediction ensures that air quality policies are both timely and effective.

Urban planning benefits greatly from satellite-integrated air quality analysis. Remote sensing identifies pollution corridors associated with traffic congestion, industrial zones, and dense urban development. Planners can use this data to redesign road networks, relocate polluting activities, and improve ventilation through green infrastructure. Satellite observations also help evaluate the effectiveness of urban design changes over time. By continuously monitoring air quality, cities can refine strategies and adapt to population growth. Space technology enables planners to design healthier cities based on evidence rather than assumptions.

AI-enhanced satellite monitoring improves understanding of pollution's interaction with weather and climate. Atmospheric conditions such as temperature inversions, wind speed, and humidity influence how pollutants accumulate and disperse. Satellites capture these variables at large scales, while AI models analyze their combined effects on air quality. This integrated approach helps explain sudden pollution events and seasonal patterns. By understanding these interactions, policymakers can tailor interventions to local environmental conditions. Space-based insights ensure that air quality management accounts for both human activity and natural atmospheric processes.

Public access to satellite-derived air quality information empowers communities and strengthens accountability. Many governments and organizations publish satellite-based pollution maps and forecasts online. Citizens can make informed decisions about outdoor activities, transportation choices, and health precautions. Transparency also encourages industries and municipalities to comply with environmental regulations. By democratizing access to air quality data, space technology fosters public engagement and trust. Informed communities become active participants in creating cleaner, healthier environments.

Satellite air quality monitoring plays a critical role in regulatory enforcement and environmental compliance. Governments use space-based data to verify emissions from industrial facilities, power plants, and transportation corridors. Satellites detect pollution levels independent of self-reported data, increasing transparency and accountability. When emission thresholds are exceeded, authorities can initiate inspections, enforce penalties, or require corrective actions. This objective monitoring discourages non-compliance and strengthens environmental governance. Space technology ensures that air quality regulations are enforced consistently and fairly.

International cooperation is enhanced through shared satellite air quality data. Air pollution often crosses national borders, making it a regional or global issue rather than a local one. Satellites provide a unified view of pollutant transport, allowing neighboring countries to identify shared sources and impacts. This information supports collaborative agreements and coordinated emission reduction strategies. By relying on neutral, space-based data, nations can engage in constructive dialogue grounded in scientific evidence. Space technology fosters trust and cooperation in addressing transboundary air pollution.

Satellite monitoring supports global climate and environmental agreements by tracking greenhouse gas emissions alongside air pollutants. Space-based sensors measure carbon dioxide and methane concentrations, helping countries assess progress toward international climate commitments. This data enables verification of emission reductions and highlights areas requiring additional action. Integrating air quality and climate monitoring creates a comprehensive framework for environmental policy. Governments can align pollution control measures with climate mitigation strategies for maximum impact. Space technology ensures that global environmental goals are measurable and achievable.

Regulatory agencies use satellite data to prioritize interventions and optimize enforcement resources. By identifying pollution hotspots and recurring violations, authorities can focus inspections where they are most needed. This targeted approach reduces costs while improving compliance outcomes. Satellite monitoring also supports long-term policy evaluation by revealing trends over time. Policymakers can assess which regulations are effective and adjust strategies accordingly. Space-enabled enforcement enhances efficiency, effectiveness, and accountability in air quality management.

Public reporting and international transparency are strengthened through satellite-based air quality monitoring. Global platforms share pollution data openly, allowing researchers, journalists, and citizens to track environmental performance. This visibility pressures governments and industries to meet air quality standards and adopt cleaner practices. Public access to data also supports environmental justice by highlighting disparities in pollution exposure. Space technology empowers societies to demand cleaner air and responsible governance. Transparency driven by satellite data becomes a catalyst for lasting environmental improvement.

Space-based air quality monitoring has fundamentally transformed how societies understand, manage, and improve the air people breathe. Satellites provide continuous, objective, and large-scale observations of pollutants that affect human health and the environment. This capability allows governments, scientists, and communities to detect pollution sources, track trends, and evaluate the effectiveness of interventions. By integrating satellite data with ground measurements and predictive models, air quality management becomes more accurate and responsive. Space technology ensures that decisions about environmental protection are informed by reliable, real-time evidence. Cleaner air becomes an achievable goal rather than an abstract aspiration.

Public health protection is one of the most significant benefits of satellite air quality monitoring. By identifying areas of high pollution exposure, health authorities can issue warnings, prepare healthcare systems, and guide vulnerable populations. Long-term satellite data also supports research linking air pollution to chronic diseases, strengthening preventive health strategies. Urban planners use these insights to design healthier cities with improved transportation systems and green infrastructure. Space-based monitoring ensures that health-focused policies are proactive rather than reactive. Protecting human well-being becomes an integral part of environmental management.

Policy effectiveness and accountability are strengthened through independent satellite observations. Governments can verify emission reductions, enforce regulations, and measure progress toward air quality goals. Transparent data builds public trust and encourages compliance from industries and municipalities. Satellite monitoring also supports international agreements by providing consistent metrics across borders. By grounding policy in scientific observation, decision-makers can implement more effective and equitable solutions. Space technology reinforces governance systems that prioritize environmental and public health outcomes.

Global collaboration is made possible through shared satellite air quality data. Countries can address transboundary pollution collectively, aligning strategies to reduce emissions and protect shared ecosystems. International organizations rely on satellite data to coordinate responses to pollution events and track global trends. This shared perspective promotes cooperation rather than conflict, emphasizing collective responsibility for the atmosphere. Space technology enables a unified approach to managing a resource that knows no borders. Cleaner air becomes a shared global objective supported by common data and understanding.

Ultimately, air quality monitoring with space-based sensors demonstrates how space technology directly improves life on Earth. By providing accurate, timely, and comprehensive data, satellites empower societies to reduce pollution, protect health, and promote sustainable development. The integration of observation, analytics, policy, and public engagement creates a powerful framework for environmental action. Space technology bridges science and society, turning information into meaningful change. Through continued investment in satellite monitoring, humanity moves closer to a future where clean air is a universal right rather than a privilege.

Chapter . 18

Disaster Response Coordination from Orbit

Leveraging satellite networks to coordinate relief efforts more efficiently after disasters.

Natural disasters pose immediate and complex challenges that demand rapid, coordinated responses, and satellite technology has become essential in managing these crises. When earthquakes, hurricanes, floods, or wildfires strike, ground-based communication and infrastructure are often damaged or destroyed. Satellites provide uninterrupted communication, real-time imagery, and situational awareness that are critical during the first hours of a disaster. By observing affected areas from orbit, responders gain a comprehensive understanding of damage extent, accessibility, and priority needs. Space-based coordination ensures that emergency efforts are informed, timely, and effective. Satellites transform disaster response from fragmented action into organized, data-driven intervention.

Satellite imagery plays a central role in assessing damage immediately after a disaster. High-resolution images reveal collapsed buildings, flooded regions, burned landscapes, and blocked transportation routes. Emergency managers use this information to identify the most severely affected areas and prioritize rescue operations. Rapid damage assessment reduces response time and prevents duplication of efforts. Satellites also enable before-and-after comparisons, helping responders understand how conditions have changed. This clarity allows relief agencies to act decisively and efficiently during critical moments.

Communication satellites are vital when terrestrial networks fail during disasters. Emergency responders rely on satellite phones, internet links, and navigation systems to coordinate search and rescue operations. These systems enable real-time communication between local responders, national agencies, and international aid organizations. Reliable communication ensures that information flows quickly, reducing confusion and delays. Satellite connectivity also supports command centers that manage logistics, personnel deployment, and medical response. Space-based communication networks keep responders connected when it matters most.

Navigation satellites support precise positioning and movement during disaster response. Global navigation satellite systems guide rescue teams through damaged or unfamiliar terrain. Accurate positioning is essential for locating survivors, delivering supplies, and coordinating air and ground operations. Satellites also support drone and aircraft operations used in search and rescue missions. By providing accurate location data, space technology improves safety and efficiency for responders working in hazardous conditions. Navigation from orbit ensures that every movement is deliberate and effective.

Beyond immediate response, satellites support disaster preparedness and early warning systems. Earth observation satellites monitor weather patterns, seismic activity, and environmental changes that signal impending disasters. Early warnings allow authorities to evacuate populations, secure infrastructure, and pre-position supplies. This proactive use of satellite data reduces casualties and economic losses. By integrating early warning with response coordination, space technology creates a continuous disaster management cycle. Satellites help societies move from crisis reaction to resilience planning.

Real-world disaster responses demonstrate the life-saving impact of satellite coordination. During major earthquakes, satellites rapidly assess structural damage across affected cities. High-resolution imagery identifies collapsed buildings, blocked roads, and areas with trapped survivors. Emergency responders use this information to prioritize search and rescue missions and allocate personnel efficiently. Satellite-guided assessments reduce uncertainty during chaotic situations and help responders focus on the most critical needs. Space technology ensures that limited time and resources are used where they can save the most lives.

Hurricanes and typhoons highlight the importance of satellite monitoring before, during, and after landfall. Satellites track storm formation, intensity, and movement days in advance, enabling early warnings and evacuations. After impact, satellite imagery reveals flooded neighborhoods, damaged infrastructure, and isolated communities. Relief agencies rely on this data to plan evacuation routes, deliver supplies, and restore essential services. Continuous satellite observation allows authorities to adapt response strategies as conditions evolve. Space-based coordination minimizes confusion and accelerates recovery efforts in storm-affected regions.

Flood response has been significantly improved through satellite observation and coordination. Satellites monitor rainfall, river levels, and floodplain expansion in real time. During major floods, imagery identifies submerged roads, bridges, and settlements that may be inaccessible by ground teams. This information helps responders deploy boats, helicopters, and relief supplies efficiently. Satellite data also supports early warnings, giving communities time to evacuate and prepare. Space technology turns flood response into a proactive and well-coordinated operation.

Wildfire management benefits greatly from satellite-based disaster coordination. Satellites detect fire ignition points, track fire spread, and monitor smoke plumes over large areas. Firefighters use this data to position crews, plan containment strategies, and protect critical infrastructure. Satellite monitoring also supports evacuation planning by identifying areas at immediate risk. Real-time updates allow responders to adapt to rapidly changing fire behavior. Space technology enhances situational awareness and improves safety for both responders and civilians.

Humanitarian crises following disasters are also addressed more effectively through satellite coordination. After earthquakes, floods, or conflicts, satellites assess population displacement, shelter availability, and access to essential resources. Aid organizations use this information to plan camps, distribute food and water, and restore healthcare services. Satellite communication ensures coordination among multiple agencies working in challenging environments. By providing a shared operational picture, space technology reduces duplication and gaps in aid delivery. Satellites ensure that humanitarian assistance reaches those who need it most.

Integrating satellite data with artificial intelligence significantly enhances disaster response coordination. AI algorithms rapidly analyze satellite imagery to identify damage patterns, flooded zones, collapsed structures, and active fire fronts. This automation reduces the time required to interpret complex data during emergencies. By highlighting priority areas, AI enables responders to focus on life-saving operations rather than data processing. Machine learning models also improve accuracy over time by learning from past disasters. Space technology combined with AI accelerates decision-making when every minute matters.

Drones and unmanned aerial systems complement satellite observations by providing detailed, close-range views of affected areas. Satellites identify broad damage zones, while drones capture high-resolution imagery for specific locations. This layered approach allows responders to verify satellite assessments and gain situational awareness at multiple scales. Drones can access areas that are dangerous or inaccessible to ground teams. Coordinated use of satellites and drones ensures comprehensive coverage during disaster response. Space-based coordination extends seamlessly from orbit to the ground.

Ground systems and emergency operations centers rely on satellite data for real-time coordination. Satellite imagery, weather updates, and communication links feed into centralized command platforms. Decision-makers use this information to deploy personnel, manage logistics, and track mission progress. Integrated systems ensure that responders from different agencies share a common operational picture. This coordination reduces miscommunication and improves efficiency. Space technology enables unified command and control during complex disaster responses.

Predictive modeling enhances disaster preparedness and response when combined with satellite data. AI models use satellite observations and historical data to forecast disaster impacts and secondary hazards. For example, models can predict landslides after heavy rainfall or disease outbreaks following floods. These forecasts allow responders to plan ahead and allocate resources strategically. Predictive insights also guide evacuation planning and infrastructure protection. Space-enabled prediction transforms disaster response into a proactive and adaptive process.

Data sharing and interoperability are critical components of satellite-enabled disaster coordination. International agencies, governments, and humanitarian organizations rely on standardized satellite data formats to collaborate effectively. Shared platforms allow multiple stakeholders to access real-time information simultaneously. This transparency supports coordinated action and reduces duplication of efforts. Space technology fosters collaboration across borders and organizations during global emergencies. Satellites provide the common data foundation needed for collective disaster response.

Satellite technology plays a critical role in managing logistics and supply chains during disaster response and recovery. After a disaster, transportation networks are often disrupted, making it difficult to deliver food, water, medicine, and shelter. Satellite imagery identifies damaged roads, bridges, ports, and airports, helping logistics teams plan alternative routes. Real-time monitoring ensures that supplies reach affected areas as efficiently as possible. By reducing delays and uncertainty, satellite-supported logistics save lives and prevent secondary crises. Space technology ensures that relief operations remain functional even when ground infrastructure fails.

Supply chain coordination benefits from satellite communication and tracking systems. Relief agencies use satellite links to coordinate shipments, monitor inventory levels, and communicate with field teams. Navigation satellites track the movement of vehicles, aircraft, and ships carrying aid supplies. This visibility helps prevent bottlenecks, duplication, and misallocation of resources. Satellite-enabled logistics systems ensure that assistance is delivered where it is needed most. Space technology transforms disaster logistics into a transparent and accountable process.

Infrastructure restoration is accelerated through satellite monitoring and assessment. Satellites assess damage to power grids, water systems, telecommunications networks, and transportation corridors. Engineers use this information to prioritize repairs and allocate technical teams efficiently. Continuous satellite observation tracks progress and identifies emerging risks, such as unstable terrain or secondary flooding. This supports safe and systematic reconstruction efforts. Space technology enables faster recovery and reduces long-term disruption for affected communities.

Energy and utility restoration also rely on satellite data for coordination. Power outages following disasters can last for weeks without accurate situational awareness. Satellites identify damaged transmission lines, substations, and generation facilities across large regions. Utility companies use this data to plan repair sequences and restore service strategically. Satellite communication ensures coordination between field crews and control centers. Space-based insights help restore essential services quickly and safely.

Long-term recovery planning benefits from satellite-supported analysis of environmental and social impacts. Satellites monitor land changes, population displacement, and ecosystem damage after disasters. This data informs rebuilding strategies that reduce future vulnerability and improve resilience. Governments and organizations use satellite insights to redesign infrastructure, improve land-use planning, and strengthen disaster preparedness. Space technology ensures that recovery efforts do not simply rebuild what was lost but create safer and more resilient communities. Satellites support a transition from emergency response to sustainable recovery.

Satellite-enabled disaster response coordination has fundamentally improved humanity's ability to respond to crises quickly and effectively. By providing real-time imagery, communication, and navigation, satellites ensure that responders operate with clarity even in chaotic conditions. Damage assessment, search and rescue, logistics, and infrastructure restoration all benefit from space-based insights. These capabilities reduce response times, save lives, and minimize economic losses. Space technology transforms disaster response from fragmented action into coordinated strategy. The ability to see, communicate, and act from orbit is now a cornerstone of modern emergency management.

The integration of satellites with AI, drones, and ground systems further strengthens disaster resilience. Automated analysis accelerates decision-making, while layered observation ensures comprehensive situational awareness. Predictive models enable proactive planning, reducing the impact of disasters before they occur. This integrated approach allows responders to anticipate challenges and adapt strategies in real time. Space technology supports a continuous disaster management cycle that includes preparedness, response, recovery, and mitigation. By embracing these tools, societies become more resilient to future shocks.

Disaster recovery and long-term resilience are enhanced through satellite monitoring and planning. Continuous observation informs infrastructure rebuilding, land-use decisions, and environmental restoration. Satellites help ensure that recovery efforts reduce vulnerability rather than recreate risk. Data-driven reconstruction strengthens communities and supports sustainable development. Space-based insights guide investment in safer infrastructure and better preparedness systems. Disaster recovery becomes an opportunity to build back stronger and smarter.

Global cooperation is strengthened through shared satellite data and coordinated response efforts. Disasters often exceed the capacity of any single nation, requiring international support. Satellite networks provide a common operational picture that enables collaboration among governments, humanitarian organizations, and military responders. Shared data reduces duplication, improves efficiency, and ensures aid reaches those most in need. Space technology fosters solidarity and collective responsibility during global emergencies. Satellites serve as a unifying platform for humanitarian action.

Ultimately, disaster response coordination from orbit demonstrates how space technology directly safeguards lives and livelihoods on Earth. Satellites empower responders with knowledge, connectivity, and foresight during humanity's most vulnerable moments. By integrating observation, communication, and prediction, space-based systems create faster, fairer, and more effective disaster management. Continued investment in satellite infrastructure strengthens global resilience in an era of increasing climate and environmental risk. Space technology ensures that when disasters strike, humanity is better prepared to respond together.

Chapter . 19

Protecting Critical Infrastructure Using Space Insights

Using space technology to monitor and safeguard bridges, dams, power grids, and essential systems.

Critical infrastructure forms the backbone of modern society, supporting transportation, energy, water, communication, and economic stability. Damage or failure of these systems can have cascading effects on public safety, health, and national security. Space technology provides a powerful means to monitor infrastructure conditions continuously and at scale. Satellites observe structural movement, environmental stress, and surrounding land changes that may threaten infrastructure integrity. By detecting risks early, authorities can prevent failures before they become disasters. Space-based monitoring transforms infrastructure protection from reactive repair to proactive prevention.

Satellite imagery enables continuous surveillance of large and remote infrastructure networks. Bridges, dams, pipelines, and power transmission lines often span vast areas that are difficult to inspect regularly from the ground. Satellites provide frequent, high-resolution observations that reveal subtle changes over time. These observations help identify cracks, deformation, land subsidence, or erosion that could compromise structural safety. Early detection allows engineers to schedule maintenance and repairs before failures occur. Space technology ensures that infrastructure health is monitored consistently and comprehensively.

Interferometric synthetic aperture radar, or InSAR, is a key satellite technology for infrastructure monitoring. InSAR detects millimeter-level ground movement by comparing radar images taken over time. This capability is especially valuable for monitoring bridges, dams, railways, and buildings in earthquake-prone or subsiding regions. Engineers use InSAR data to assess structural stability and identify areas at risk of collapse. Continuous monitoring reduces uncertainty and improves safety planning. Space-based radar provides precise, reliable insights that ground inspections alone cannot achieve.

Environmental factors such as flooding, landslides, and extreme weather pose significant risks to infrastructure. Satellites monitor rainfall, soil moisture, river levels, and vegetation cover that influence these hazards. By correlating environmental data with infrastructure locations, authorities can anticipate stress and potential damage. This enables preventive measures such as reinforcing structures, adjusting operations, or issuing warnings. Space technology helps manage the growing risks posed by climate change to critical systems. Infrastructure protection becomes more adaptive and resilient through satellite observation.

Energy and communication networks rely heavily on satellite monitoring for security and reliability. Satellites track power grid performance, detect vegetation encroachment near transmission lines, and monitor pipeline routes for leaks or illegal activity. Communication satellites ensure continuity of service even when ground infrastructure is compromised. Space-based monitoring enhances situational awareness and supports rapid response to disruptions. By safeguarding essential services, space technology protects economic stability and public safety. Infrastructure resilience is strengthened through constant vigilance from orbit.

Real-world case studies illustrate how satellite monitoring has prevented infrastructure failures and reduced risk. In Italy, satellite-based InSAR monitoring has been used to detect gradual ground subsidence beneath historic bridges and urban infrastructure. Engineers identified early signs of structural stress and implemented reinforcement measures before major damage occurred. This proactive intervention preserved public safety and reduced costly emergency repairs. Satellite insights provided continuous oversight without disrupting daily operations. Space technology enabled authorities to act early rather than respond after failure.

In California, satellites play a critical role in monitoring dams and reservoirs in earthquake-prone regions. InSAR data detects subtle shifts in dam walls and surrounding terrain that may indicate structural weakening. Combined with hydrological satellite data tracking water levels and rainfall, authorities can manage stress on these structures more effectively. Early warnings allow operators to release water safely or reinforce vulnerable sections. This integrated monitoring reduces the risk of catastrophic dam failure. Space-based observation strengthens water security and disaster preparedness.

Railway networks across Europe rely on satellite monitoring to ensure track stability and safety. Satellites detect land movement, erosion, and subsidence along rail corridors that could compromise track alignment. Transportation authorities use this information to prioritize inspections and maintenance schedules. Early detection prevents derailments and service disruptions, protecting passengers and freight operations. Satellite monitoring also reduces inspection costs by focusing resources where they are most needed. Space technology enhances both safety and efficiency in transportation infrastructure management.

In Southeast Asia, satellite monitoring has been essential in protecting levees and flood control systems. Satellites track river levels, soil saturation, and structural deformation along levee networks. Authorities use this data to identify weak points and reinforce them before flooding occurs. Early intervention prevents breaches that could devastate communities and agriculture. Continuous satellite surveillance ensures that flood protection systems remain reliable under changing climate conditions. Space technology supports resilient water management and disaster prevention.

Energy infrastructure protection also benefits from satellite insights. In Canada, satellites monitor long-distance oil and gas pipelines for ground movement, landslides, and unauthorized activity. InSAR detects deformation that may indicate stress or potential rupture. Operators can respond quickly by adjusting operations or conducting targeted inspections. Satellite monitoring reduces environmental risks and enhances operational safety. Space technology provides a critical layer of protection for energy systems that support modern economies.

Integrating satellite data with artificial intelligence and predictive analytics enhances infrastructure resilience and maintenance planning. AI algorithms analyze large volumes of satellite imagery to detect subtle changes that may indicate structural weakness or environmental stress. Predictive models forecast potential failures based on historical data, seasonal patterns, and observed deformation. This allows engineers to prioritize repairs, reinforce structures, and mitigate risks before incidents occur. By combining satellite observation with AI, infrastructure management becomes proactive rather than reactive. Space technology ensures that infrastructure systems remain safe, reliable, and resilient over time.

Digital twins, or virtual replicas of infrastructure, benefit from satellite-derived data for continuous monitoring and simulation. Satellites provide high-resolution imagery and geospatial data that feed into digital models of bridges, dams, pipelines, and power grids. Engineers can simulate stress scenarios, assess vulnerability, and evaluate maintenance strategies. This enables predictive maintenance and optimized resource allocation. Satellite data ensures that digital twins reflect real-world conditions accurately, even for remote or inaccessible infrastructure. Space technology transforms infrastructure management into a data-driven, intelligent process.

Transportation systems gain improved reliability and safety through satellite monitoring. Highways, railways, ports, and airports are vulnerable to environmental stressors such as landslides, floods, or earthquakes. Satellites detect early warning signs and track changes that may compromise operational safety.

Predictive models integrate this data to optimize maintenance schedules and minimize disruptions. Real-time monitoring ensures rapid response to emerging threats, protecting both passengers and cargo. Space technology provides the foundation for resilient, adaptive, and efficient transportation networks.

Energy grids are also strengthened through satellite-based insights. Power transmission lines, substations, and renewable energy installations face risks from extreme weather, ground movement, and vegetation encroachment. Satellites monitor these risks continuously, allowing operators to prevent outages or infrastructure damage. Predictive analytics optimize maintenance planning, reducing downtime and operational costs. Space-enabled monitoring ensures that energy systems remain reliable, secure, and resilient against environmental and technical hazards. Satellite technology bridges observation and action for critical energy infrastructure.

Urban infrastructure benefits from integrating satellite data with predictive modeling and smart city planning. Buildings, water networks, and transportation systems can be assessed for vulnerabilities due to earthquakes, flooding, or subsidence. Satellite observations provide accurate, up-to-date information for zoning, construction standards, and emergency planning. Predictive analytics allow city planners to prioritize retrofitting, strengthening, or relocation of at-risk structures. Space technology ensures that urban infrastructure remains safe, efficient, and capable of supporting growing populations. Satellites transform urban resilience into a measurable and manageable objective.

Satellite data, AI, and predictive modeling have already prevented significant infrastructure failures in multiple countries. In Japan, space-based monitoring has been crucial for earthquake-prone areas. Satellites detect subtle ground shifts and structural stress in bridges, buildings, and dams, allowing engineers to reinforce infrastructure before earthquakes cause major damage. Predictive models integrate historical earthquake data with satellite observations to forecast potential high-risk zones. This proactive approach reduces casualties and economic losses. Space technology transforms earthquake preparedness from reactive response to preventive action.

In the Netherlands, satellite monitoring supports flood defense systems that protect densely populated regions below sea level. Dikes, levees, and storm surge barriers are continuously monitored for deformation, soil erosion, and water pressure. AI models predict weaknesses or failure points, enabling timely repairs or reinforcement. Satellite observation ensures that critical water management infrastructure remains robust against rising sea levels and extreme weather events. Space-enabled monitoring strengthens national resilience and safeguards communities in vulnerable areas.

Energy infrastructure in Europe benefits from predictive satellite monitoring. High-voltage power lines crossing forests and mountainous regions are at risk from landslides, heavy snow, or vegetation encroachment. Satellites detect potential hazards before they cause outages or damage. Predictive analytics guide preventive maintenance and risk mitigation strategies. Operators can deploy crews efficiently, minimizing downtime and preventing costly disruptions. Space technology ensures a reliable energy supply while reducing the risk of accidents.

In South America, satellite monitoring helps protect transportation infrastructure from landslides and riverbank erosion. Roads, bridges, and railways in mountainous or flood-prone areas are continuously observed. AI algorithms analyze terrain changes, rainfall patterns, and soil stability to predict areas at risk. Emergency interventions, such as reinforcing embankments or rerouting traffic, can be implemented before catastrophic failures occur. Satellite-supported monitoring saves lives, protects economic activity, and ensures continuity of essential services. Space technology provides early warnings that make proactive action possible.

Satellite monitoring also strengthens critical infrastructure security against human threats. Pipelines, power grids, and communication networks can be monitored for unauthorized activity, sabotage, or cyber-physical attacks. Satellites detect anomalies in operational patterns, environmental interference, or unusual access points. Combined with predictive analytics, authorities can respond swiftly to threats, preventing damage and maintaining service continuity. Space technology enhances both the safety and resilience of vital infrastructure systems. Satellites ensure that critical assets remain secure from natural and human-induced risks.

Space-based monitoring has become an essential tool in safeguarding critical infrastructure and protecting public safety. Satellites provide continuous observation of bridges, dams, power grids, pipelines, and transportation networks, enabling early detection of structural weaknesses and environmental threats. By combining satellite data with AI, predictive modeling, and digital twins, authorities can forecast risks, plan preventive maintenance, and respond before failures occur. This proactive approach reduces the likelihood of disasters, protects lives, and preserves economic stability. Space technology transforms infrastructure management from reactive repair to strategic resilience planning.

Disaster resilience is enhanced when satellites are integrated into national infrastructure monitoring programs. Continuous monitoring ensures that emerging threats from natural hazards, climate change, or human activity are detected promptly. Predictive insights guide policymakers, engineers, and emergency responders in prioritizing interventions, allocating resources, and implementing preventive measures. Space-based data supports informed decision-making, reduces uncertainties, and strengthens the capacity to maintain critical services during emergencies. Satellites enable infrastructure systems to withstand shocks while minimizing social and economic disruption.

Energy and transportation networks benefit significantly from satellite surveillance. Power lines, substations, roads, and railways are continuously observed to detect deformation, environmental stress, or unauthorized interference. Early warnings allow rapid response, reducing downtime and preventing cascading failures. Satellite-enabled monitoring ensures the reliability and security of essential services. By maintaining operational continuity, space technology protects communities, businesses, and national economies. Critical systems are safeguarded against both natural and human-induced risks.

Urban infrastructure and public facilities also gain from space-based insights. Hospitals, water systems, bridges, and emergency shelters are monitored for structural stability and environmental threats. Predictive models using satellite data support city planning, maintenance scheduling, and disaster preparedness. Continuous observation ensures that urban infrastructure is resilient, efficient, and capable of serving growing populations. Space technology provides cities with the tools to protect residents and maintain essential services under stress. Satellites transform urban resilience from theoretical planning into actionable intelligence.

Ultimately, protecting critical infrastructure with space insights demonstrates how satellite technology directly improves life on Earth. By combining observation, analytics, and predictive planning, satellites enable proactive risk management, enhance disaster resilience, and safeguard public safety. Governments, industries, and communities can respond faster, make informed decisions, and minimize the impact of natural and human-induced hazards. Space technology ensures that infrastructure systems remain secure, reliable, and sustainable. Satellites provide the intelligence needed to protect society's most vital assets today and for the future.

Chapter . 20

Space Technology in Humanitarian Aid Delivery

Applying satellite data to ensure aid reaches the most vulnerable populations effectively.

Humanitarian crises caused by natural disasters, conflicts, or pandemics demand rapid and precise delivery of aid to affected populations. Space technology has become an indispensable tool for coordinating these efforts. Satellites provide real-time imagery, mapping, and communication capabilities that allow organizations to identify communities in need, plan delivery routes, and monitor aid distribution. By observing terrain, road conditions, and damaged infrastructure from orbit, satellites enable aid agencies to optimize logistics and prioritize areas most at risk. Space-based data ensures that assistance reaches vulnerable populations efficiently, reducing delays and preventing shortages. Humanitarian aid delivery is transformed from reactive response to proactive, data-driven intervention through space technology.

Satellite imagery helps map affected regions accurately, especially in remote or conflict-affected areas. High-resolution images identify destroyed infrastructure, inaccessible roads, flooded zones, and isolated communities. This information guides relief organizations in planning safe and effective delivery routes for food, water, medicine, and shelter. Satellites also provide situational updates after initial assessments, allowing responders to adjust plans as conditions change. By combining spatial intelligence with ground reports, aid agencies gain a comprehensive understanding of the humanitarian landscape. Space technology provides a strategic overview that enhances operational decision-making.

Communication satellites are essential when terrestrial networks are damaged or non-existent. They enable coordination between field teams, command centers, and international organizations. Satellite communication ensures that updates on supply status, delivery progress, and emerging needs reach all stakeholders in real time. This connectivity reduces errors, duplication, and delays in aid distribution. Reliable communication networks also support telemedicine, remote assessment, and coordination of emergency shelters. Space-based communication ensures continuity of operations even under the most challenging conditions.

Predictive modeling, informed by satellite data, enhances humanitarian planning and risk assessment. Satellites monitor environmental variables such as rainfall, river levels, soil stability, and vegetation cover, which influence the likelihood of disasters or secondary crises. AI algorithms use these data to forecast regions likely to face shortages, disease outbreaks, or infrastructure collapse. Humanitarian organizations can pre-position supplies, deploy personnel, and issue alerts to vulnerable communities. Satellite-assisted predictive models increase efficiency, reduce waste, and save lives. Space technology transforms humanitarian aid into a proactive, strategic endeavor.

Satellite monitoring also supports transparency, accountability, and evaluation in aid delivery. High-resolution imagery allows organizations and donors to verify whether resources reach intended recipients. Monitoring can detect gaps, misallocation, or diversion of aid, enabling corrective action. Continuous observation ensures that aid efforts remain focused on the most critical needs and maintain donor and public trust. By providing an objective and verifiable record, satellites enhance governance and ethical standards in humanitarian response. Space technology ensures that assistance is both effective and equitable.

Several real-world humanitarian operations demonstrate the critical role of satellite technology in delivering aid effectively. During the 2010 Haiti earthquake, satellites provided rapid assessment of collapsed buildings, blocked roads, and damaged infrastructure. Relief agencies used this imagery to prioritize aid delivery to the hardest-hit areas. Satellite mapping allowed helicopters and convoys to navigate around rubble and debris efficiently. By identifying accessible routes and population clusters in need, satellites minimized delays and enhanced the effectiveness of aid distribution. Space technology ensured that life-saving assistance reached affected communities quickly.

In East Africa, satellite monitoring has been used to coordinate famine relief efforts. Drought conditions, crop failures, and water scarcity are tracked using satellite imagery of vegetation health, soil moisture, and rainfall patterns. Early detection of food insecurity allows humanitarian agencies to pre-position food, water, and medical supplies. Satellite-assisted monitoring ensures that remote and vulnerable populations receive timely assistance before shortages become catastrophic. By providing accurate, real-time environmental data, satellites support evidence-based decision-making in humanitarian logistics. Space technology transforms famine relief from reactive intervention into proactive prevention.

During the Syrian conflict, satellite imagery played a critical role in delivering aid to populations in inaccessible or dangerous areas. Satellites identified areas of destruction, displaced populations, and potential delivery routes that were safe for humanitarian convoys. This information guided NGOs and international organizations in reaching communities cut off by conflict. Satellite communication networks enabled coordination between multiple organizations across borders. By providing reliable situational awareness, satellites ensured that aid reached civilians despite complex logistical and security challenges. Space technology bridges the gap between intention and action in conflict zones.

The COVID-19 pandemic also highlighted the value of satellite data in global humanitarian logistics. Satellites tracked population movement patterns, urban density, and transportation networks, helping organizations plan vaccine distribution and medical supply delivery. Remote monitoring allowed authorities to prioritize high-risk regions and underserved communities. By combining satellite observation with epidemiological models, aid agencies could allocate resources efficiently and reduce disease spread. Space technology enhanced coordination, transparency, and effectiveness in pandemic response. Satellites ensured that health and humanitarian aid reached the most vulnerable populations.

In addition, satellite monitoring supports disaster preparedness for humanitarian operations. Early detection of hurricanes, floods, and cyclones allows aid agencies to pre-position supplies, deploy teams, and prepare shelters. Satellite forecasts inform evacuation planning and logistics before disasters strike. By anticipating needs and vulnerabilities, humanitarian organizations reduce response times and improve outcomes. Space technology enables a shift from reactive to anticipatory aid delivery. Satellites empower humanitarian actors to save lives and reduce suffering during crises.

Integrating satellite data with artificial intelligence enhances the precision and efficiency of humanitarian aid delivery. AI algorithms analyze satellite imagery to detect the location of affected populations, damaged infrastructure, and inaccessible routes. This analysis allows relief agencies to prioritize areas in greatest need and optimize the allocation of resources. Predictive models forecast how disasters will impact communities over time, guiding proactive interventions. By combining satellite observation with AI, aid delivery becomes faster, smarter, and more targeted. Space technology ensures that humanitarian efforts are both efficient and effective.

Drones complement satellite monitoring by providing high-resolution imagery of specific areas identified from orbit. Satellites identify broad damage zones, while drones capture detailed information for precise planning. This layered approach allows field teams to navigate difficult terrain, assess damage, and deliver supplies more effectively. Drones can also transport small emergency packages, such as medical kits or vaccines, directly to inaccessible locations. Integration with satellite data ensures that drones operate efficiently and safely. Space technology creates a seamless connection between macro-level monitoring and micro-level response.

Mobile applications connected to satellite networks improve communication and coordination in humanitarian operations. Field teams, local volunteers, and aid agencies can share real-time updates on supply status, population needs, and transportation conditions. Satellite communication ensures that mobile apps function even in areas where terrestrial networks are damaged or nonexistent. This connectivity allows organizations to adjust plans rapidly, reducing delays and duplication of effort. By linking satellites, AI, drones, and mobile technology, humanitarian operations become highly adaptive. Space-enabled coordination ensures that aid delivery responds to evolving conditions effectively.

Supply chain management in humanitarian operations is enhanced through satellite monitoring. Satellites track road conditions, weather patterns, and environmental hazards that affect transportation routes. Predictive analytics anticipate potential disruptions and optimize logistics planning. Humanitarian agencies can reroute convoys, pre-position supplies, and adjust delivery schedules based on real-time satellite data. This approach reduces wasted resources, prevents delays, and ensures that critical aid reaches the intended populations. Space technology strengthens the reliability and efficiency of humanitarian supply chains.

Satellite monitoring also supports evaluation and accountability in humanitarian aid delivery. High-resolution imagery allows agencies to verify that resources reach target populations and identify gaps in coverage. This transparency ensures that aid efforts are equitable, effective, and ethically managed. Donors and governments can use satellite-based reports to track outcomes and improve future planning. Continuous observation ensures that lessons learned are incorporated into long-term humanitarian strategies. Space technology transforms accountability into a measurable, verifiable process, increasing trust and efficiency.

Several case studies highlight the impact of integrating satellites, AI, drones, and mobile applications in humanitarian aid delivery. During the 2015 Nepal earthquake, satellites rapidly mapped collapsed buildings, blocked roads, and isolated villages. AI algorithms processed the imagery to identify priority zones for relief distribution. Drones were deployed to provide high-resolution reconnaissance of areas too dangerous for ground teams. Mobile applications allowed field responders to report conditions in real time, enabling coordinated aid delivery. The combination of these technologies ensured that life-saving resources reached affected communities efficiently and effectively.

In the Philippines, satellite imagery combined with AI models played a crucial role in disaster response during typhoons. Satellites tracked storm intensity and flooding, while predictive algorithms forecasted regions at highest risk. Drones surveyed damaged areas, mapping impassable roads and collapsed infrastructure. Humanitarian teams used mobile apps to coordinate logistics, communicate with field staff, and update relief recipients. This integrated approach minimized delays and prevented duplication of efforts. Space technology provided comprehensive situational awareness that improved the speed and accuracy of aid delivery.

During the Syrian refugee crisis, satellite monitoring helped identify densely populated displacement camps and vulnerable settlements. High-resolution imagery was analyzed using AI to map shelter locations, water sources, and sanitation facilities. Drones supplemented this data by assessing conditions within camps and identifying areas in need of urgent aid. Mobile communication tools allowed NGOs to coordinate operations across multiple organizations, ensuring resources were allocated efficiently. Satellites enabled humanitarian teams to reach populations in politically sensitive or conflict-prone regions safely. Space technology bridged gaps where traditional monitoring methods were impractical.

In East Africa, drought and famine relief benefited from satellite-supported predictive models. Satellites monitored vegetation health, soil moisture, and water availability to anticipate areas of food insecurity. AI models forecasted the timing and severity of potential shortages. Drones were used to map remote rural areas and deliver emergency supplies, while mobile apps coordinated distribution among dispersed communities. This proactive approach allowed humanitarian agencies to intervene before crises became severe. Space technology ensured timely, targeted, and life-saving aid for vulnerable populations.

Satellite-based humanitarian coordination was also critical during the COVID-19 pandemic. Satellites tracked population density, mobility patterns, and urban infrastructure, guiding vaccine distribution and medical supply delivery. AI analyzed satellite data to forecast outbreak hotspots and optimize logistics. Drones assisted in last-mile delivery of vaccines and medical kits to isolated regions. Mobile applications allowed health workers to report delivery status, track recipients, and adjust distribution plans in real time. Space technology ensured equitable access to healthcare resources, especially in underserved areas. Satellites enabled humanitarian response to operate efficiently in both health and disaster contexts.

Space technology has transformed humanitarian aid delivery into a precise, efficient, and equitable process. Satellites provide real-time imagery, communication networks, and environmental data that allow relief organizations to understand affected areas comprehensively. By combining these observations with AI, drones, and mobile applications, aid delivery becomes faster, more targeted, and adaptive to evolving conditions. Vulnerable populations receive assistance when and where it is most needed, minimizing suffering and saving lives. Space technology ensures that humanitarian operations are guided by data, rather than guesswork, enhancing both efficiency and effectiveness.

Proactive disaster preparedness is strengthened through satellite monitoring and predictive modeling. Satellites track environmental indicators such as rainfall, river levels, soil conditions, and vegetation health to anticipate potential crises. AI algorithms analyze these data to forecast areas at risk of famine, flooding, or disease outbreaks. Humanitarian organizations can pre-position supplies, plan evacuation routes, and coordinate teams before disasters strike. This forward-looking approach reduces delays, prevents bottlenecks, and ensures that aid reaches communities in time. Space technology allows humanitarian action to shift from reactive response to anticipatory intervention.

Transparency and accountability in aid delivery are enhanced by satellite monitoring. High-resolution imagery verifies that resources reach intended recipients and identifies gaps in coverage or misallocation. Satellite data also supports reporting to donors, governments, and the public, fostering trust and credibility. Continuous observation allows agencies to learn from each operation, improving future strategies and resource management. Space technology ensures that humanitarian aid is delivered responsibly, ethically, and equitably. Monitoring from orbit provides an objective and verifiable record of action and impact.

Global coordination of humanitarian efforts benefits from satellite-enabled connectivity. Satellites facilitate communication between multiple agencies, governments, and international organizations operating in crisis zones. Shared satellite data creates a unified operational picture, ensuring collaboration rather than duplication or conflict. This connectivity is especially vital in remote, conflict-affected, or disaster-stricken regions where ground communication networks are compromised. Space technology strengthens international cooperation, enabling aid to be distributed efficiently across borders and among diverse organizations. Satellites foster a global network of humanitarian coordination.

Ultimately, space technology ensures that humanitarian aid reaches the people who need it most, regardless of geography or circumstance. Satellites provide the intelligence, connectivity, and foresight necessary to plan, coordinate, and execute relief operations effectively. By integrating observation, analytics, and communication, space technology transforms humanitarian action from fragmented efforts into a cohesive, data-driven system.

Communities receive faster, more precise, and more equitable assistance, while organizations maximize the impact of limited resources. Satellites demonstrate how space technology can directly improve life on Earth by saving lives, reducing suffering, and strengthening resilience.

Part V: Science, Innovation, and Education

Chapter . 21

Inspiring STEM Education Through Space Exploration

Using space achievements to motivate students in science, technology, engineering, and math.

Space exploration has long captured the imagination of students, inspiring curiosity about the universe and the principles that govern it. Achievements such as lunar missions, Mars rovers, and satellite launches provide tangible examples of science and engineering in action. These accomplishments show students that knowledge of physics, mathematics, and technology can directly influence exploration and innovation. By integrating space exploration into STEM education, educators can motivate students to pursue careers in scientific fields. Space achievements demonstrate that learning STEM is not abstract but has real-world applications that can change the world. Inspiring students through space exploration fosters the next generation of scientists, engineers, and innovators.

Hands-on learning experiences using space-themed projects engage students and deepen understanding. Activities such as building model rockets, coding simulations of space missions, and analyzing satellite imagery allow students to apply theoretical knowledge in practical ways. These projects cultivate problem-solving, critical thinking, and collaboration skills. Students can experiment, observe outcomes, and refine approaches, mimicking real-world scientific processes. By connecting classroom learning to space exploration, STEM subjects become interactive and exciting rather than purely theoretical. Space-inspired education transforms curiosity into tangible skills and knowledge.

Classroom integration of satellite data enhances students' analytical and research skills. Many educational programs provide access to real satellite imagery of Earth and other planets. Students can track climate change, urban development, natural disasters, or planetary geology. By analyzing these data sets, students learn how to collect, process, and interpret scientific information. This exposure to authentic scientific data fosters both technical proficiency and a deeper appreciation for the relevance of STEM. Space technology brings real-world scientific research directly into the classroom.

Space exploration also promotes interdisciplinary learning, connecting STEM with history, ethics, and social studies. Students can examine the historical context of space programs, explore the ethical implications of space colonization, or consider the societal impact of satellite technology. These discussions encourage students to think critically about the broader consequences of scientific advancement. By linking STEM subjects to human, environmental, and ethical considerations, space-themed education promotes holistic learning. Students gain a broader perspective on how science interacts with society.

Mentorship and role models inspired by space programs play a vital role in STEM engagement. Astronauts, engineers, and scientists serve as tangible examples of what is possible through dedication, education, and curiosity. Student engagement increases when they see individuals succeeding in fields related to space exploration. Virtual talks, interactive workshops, and STEM competitions allow students to connect with professionals, ask questions, and envision their own career paths. Space exploration not only teaches technical knowledge but also inspires ambition and perseverance. It shows that the pursuit of STEM knowledge can lead to extraordinary achievements.

Numerous programs and competitions use space exploration to inspire STEM education. Initiatives such as NASA's student programs, the European Space Agency's educational outreach, and private sector space challenges provide hands-on experiences for learners of all ages. Students can design experiments for the International Space Station, simulate satellite missions, or participate in robotics competitions. These programs provide practical exposure to engineering, coding, physics, and mathematics. By engaging students in real-world challenges, space-themed initiatives make STEM subjects tangible and relevant. Space exploration programs create opportunities for students to experience problem-solving in the context of global scientific endeavors.

Competitions like the Zero Robotics challenge and international robotics tournaments allow students to design code for satellites or robotic arms. Teams compete to complete space-related tasks, often collaborating across borders and schools. These experiences foster creativity, teamwork, and technical expertise. Students gain insight into the iterative process of design, testing, and improvement, which mirrors professional scientific and engineering practices. Competitions transform learning from passive acquisition to active experimentation. Space-themed contests motivate students to innovate and develop practical solutions using STEM knowledge.

Satellite observation projects also bring STEM learning to the classroom. Many educational organizations provide access to Earth observation data for analyzing climate patterns, urban development, or natural disasters. Students can use satellite data to conduct research, create models, and test hypotheses. These projects teach data analysis, visualization, and scientific methodology. Students gain experience interpreting real-world data, connecting classroom learning with global phenomena. By analyzing satellite information, students see the tangible impact of STEM knowledge on understanding and solving real problems.

Virtual simulations of space missions provide immersive STEM experiences. Students can simulate satellite deployments, planetary exploration, or spacecraft navigation using digital platforms. These simulations teach physics, mathematics, and engineering principles in interactive ways. By experimenting in virtual environments, students learn to troubleshoot, optimize, and iterate solutions. Simulations provide a safe, scalable, and engaging way to explore complex STEM concepts. Space exploration encourages learning through exploration, trial, and discovery, which builds both competence and confidence.

Partnerships between schools and space organizations further enhance STEM learning opportunities. Collaborations enable workshops, mentorship programs, and access to professional resources. Students learn directly from engineers, scientists, and astronauts, gaining insight into career pathways and industry expectations. Partnerships also support curriculum development, ensuring that STEM lessons are aligned with practical applications and emerging technologies. Space-inspired collaboration connects education with real-world innovation. By linking students with professionals, space programs cultivate skills, inspiration, and motivation for lifelong STEM engagement.

Space-themed STEM education has a lasting impact on students' career choices and aspirations. Exposure to space missions, satellites, and exploration programs demonstrates that STEM skills can lead to exciting and meaningful careers. Students inspired by space are more likely to pursue degrees in engineering, physics, computer science, and related disciplines. Early engagement cultivates curiosity, persistence, and problem-solving abilities that are essential for scientific success. By connecting classroom learning to real-world applications, space exploration transforms abstract concepts into tangible opportunities. Space-inspired education helps build the next generation of innovators, scientists, and explorers who will tackle Earth's most pressing challenges.

Innovation is further accelerated through space-based STEM programs. Students trained in satellite design, robotics, and data analysis develop skills that are transferable across multiple industries. Projects often involve interdisciplinary problem-solving, encouraging creativity and lateral thinking. These experiences promote a mindset that is comfortable with experimentation, iteration, and technical risk-taking. Exposure to space technology challenges students to apply their knowledge in novel ways. By learning to innovate through space-focused projects, students gain confidence to pursue solutions for real-world problems. Space education cultivates a generation equipped to drive technological advancement on Earth.

Equity and inclusion in STEM are enhanced through space education initiatives. Many programs target underrepresented groups, offering mentorship, scholarships, and access to satellite technology resources. By providing opportunities to historically marginalized students, space-based programs help close the STEM achievement gap. Inclusive programs inspire students from diverse backgrounds to consider careers in science and technology. Satellite and space-focused projects provide universal appeal, engaging learners who might otherwise feel disconnected from STEM subjects. Space exploration demonstrates that STEM knowledge and careers are accessible to everyone, regardless of background or circumstance.

STEM skills gained through space education have broader societal benefits. Understanding satellites, robotics, and space systems equips students to contribute to addressing global challenges such as climate change, disaster response, and resource management. Students learn to analyze data, model systems, and propose evidence-based solutions. These capabilities have applications far beyond space exploration itself, impacting environmental policy, infrastructure planning, and public health. Space-inspired STEM education fosters civic-minded scientists and engineers capable of improving communities and societies. Satellites and space missions serve as catalysts for both learning and real-world impact.

Global initiatives have demonstrated the transformative impact of space exploration on STEM education. Programs like NASA's "STEM Engagement," the European Space Agency's "ESA Education," and the United Nations Office for Outer Space Affairs (UNOOSA) initiatives provide students worldwide with access to satellite data, virtual labs, and hands-on projects. These initiatives allow learners to participate in experiments, simulate missions, and analyze real space-based data. Students from diverse regions gain exposure to cutting-edge science and engineering while developing technical and analytical skills. International collaboration in space-themed education broadens perspectives and encourages cross-cultural problem-solving. Space exploration programs provide a global platform for inspiring the next generation of STEM professionals.

Collaborations between universities and space agencies create pathways for advanced education and career development. Students can engage in internships, research projects, and mission simulations that provide real-world experience. Working alongside scientists and engineers, learners gain insight into the practical applications of mathematics, physics, and technology. These experiences strengthen professional skills, build confidence, and clarify potential career trajectories. University partnerships also support curriculum development that integrates space science into STEM programs. By connecting education with professional opportunities, space initiatives foster both learning and workforce readiness.

International competitions further enhance STEM engagement through space-themed challenges. Contests such as CanSat, SpaceX Hyperloop competitions, and RoboCup junior space challenges motivate students to apply STEM knowledge creatively. Teams design, build, and test projects that solve problems inspired by real space missions. Participants learn collaboration, critical thinking, and iterative design processes. These competitions foster innovation and expose students to practical applications of STEM concepts. Space-inspired challenges cultivate the creativity, persistence, and problem-solving skills necessary for scientific and technological advancement.

Satellite programs also provide students with opportunities to analyze Earth and space data for societal benefit. Initiatives allow learners to study climate change, natural disasters, or environmental management using satellite imagery and data. This hands-on research fosters analytical skills, critical thinking, and evidence-based decision-making. Students gain an understanding of how STEM knowledge can address real-world problems. Space data projects demonstrate the direct relevance of STEM learning to societal challenges, motivating students to engage deeply with scientific inquiry. Satellites serve as both educational tools and inspiration for global problem-solving.

The long-term impact of space-themed STEM education extends far beyond the classroom. Students inspired by space exploration often pursue careers in science, technology, engineering, and mathematics, contributing to innovation, research, and technological advancement. The skills they develop—critical thinking, problem-solving, collaboration, and data analysis—are applicable across a wide range of industries, from aerospace and robotics to environmental science and healthcare. Exposure to space missions and satellite technology instills curiosity, perseverance, and a sense of possibility that drives lifelong learning. By connecting STEM education to real-world challenges, space exploration cultivates professionals capable of solving complex problems on Earth and beyond.

Space education also nurtures creativity and interdisciplinary thinking. Students learn to integrate physics, engineering, mathematics, computer science, and environmental science to design experiments, analyze data, and develop solutions. This holistic approach mirrors the multifaceted nature of modern scientific and technological challenges. Learners are encouraged to think innovatively and explore multiple pathways to achieve goals. By fostering creativity alongside technical expertise, space-themed STEM education equips students with the versatility required for a rapidly evolving world. Space exploration demonstrates that curiosity and ingenuity are as important as knowledge and skill.

Global collaboration in space education cultivates international perspectives and cultural competence. Programs that connect students across countries encourage teamwork, communication, and problem-solving in diverse groups. Learners develop an appreciation for global challenges, from climate change to disaster response, and see how STEM skills can contribute to solutions. By participating in international projects, students understand that scientific achievement and technological innovation are collaborative endeavors. Space exploration serves as a unifying theme that bridges cultures, inspires collective action, and promotes shared responsibility for our planet.

Societal impact is further amplified as space-inspired students contribute to innovation ecosystems. Graduates of space-focused STEM programs often enter research institutions, tech startups, or public policy roles where they apply their skills to real-world challenges. They develop technologies, design sustainable solutions, and advance scientific understanding. By cultivating a generation of problem solvers, space education ensures that future innovations address critical global issues. Satellites, space missions, and exploration initiatives act as catalysts for the creation of knowledge, technology, and sustainable solutions on Earth.

Ultimately, space exploration inspires curiosity, drives learning, and transforms education into a pathway for societal advancement. By connecting students to tangible examples of science and technology in action, satellites and space missions make STEM engaging, relevant, and attainable. Students acquire technical skills, critical thinking abilities, and a global perspective that prepare them to lead and innovate in the 21st century. Space-themed STEM education ensures that humanity's pursuit of knowledge extends beyond the cosmos to the benefit of life on Earth. Satellites and exploration serve not only to expand our understanding of space but also to cultivate a skilled, inspired, and empowered generation for the future.

Chapter . 22

Accelerating Innovation with Space-Derived Technology

How spin-offs from space research can revolutionize industries and solve Earth-based challenges.

Space exploration and research have historically driven technological innovation with applications far beyond orbit. Technologies developed for spacecraft, satellites, and space missions often find transformative uses on Earth. These spin-offs, ranging from advanced materials and imaging systems to robotics and telecommunications, accelerate innovation across industries. By leveraging space-derived technology, businesses and governments can solve pressing challenges in healthcare, energy, manufacturing, and environmental monitoring. The iterative process of designing for the extreme conditions of space cultivates solutions that are highly efficient, resilient, and adaptable. Space research thus serves as a catalyst for technological progress that benefits humanity on multiple fronts.

Satellite imaging and remote sensing technologies are prime examples of space-derived innovations applied on Earth. Originally developed for Earth observation and planetary science, these technologies now support agriculture, urban planning, disaster response, and climate monitoring. High-resolution imagery, multi-spectral data, and real-time monitoring allow organizations to optimize resource use, predict environmental risks, and enhance operational efficiency. Industries that adopt these space-derived tools can make informed decisions based on precise and timely data. By translating space observation techniques into terrestrial applications, societies gain the ability to respond to complex problems faster and more effectively. Space technology thus transforms how data-driven decisions are made.

Materials engineering from space programs has led to significant industrial advancements. Lightweight composites, thermal insulation materials, and radiation-resistant coatings were initially developed to withstand the harsh environment of space. These materials now find applications in construction, transportation, electronics, and medical devices. By meeting the extreme demands of space travel, engineers have created products that are stronger, more durable, and more energy-efficient. Industries adopt these materials to enhance performance, reduce costs, and extend product lifespans. Space-derived materials exemplify how innovations designed for extraterrestrial conditions can directly improve life on Earth.

Robotics and automation technologies developed for space missions are also accelerating industrial innovation. Robotic arms, autonomous vehicles, and remote-controlled systems were first designed for satellite servicing, planetary exploration, and space station maintenance. On Earth, these systems are adapted for manufacturing, healthcare, disaster response, and precision agriculture. Space-derived robotics improve productivity, safety, and operational precision in challenging or hazardous environments. By applying technologies proven in space, industries reduce human risk while increasing efficiency and accuracy. Space research demonstrates that extreme engineering challenges can drive practical innovations for everyday life.

Telecommunication and navigation systems are among the most influential space-derived technologies impacting daily life. Satellite communication, GPS, and Earth observation networks, initially developed for space missions, now underpin global connectivity, transportation, and commerce. These systems facilitate navigation, logistics optimization, remote learning, and telemedicine. Businesses, governments, and individuals rely on satellite-based networks for efficient, accurate, and secure communication. Space-derived technology has become essential to modern infrastructure and societal functioning. Satellites exemplify how space innovation can directly enhance global connectivity and operational capability.

Medical technology has benefited significantly from space-derived innovations. Space missions require compact, precise, and reliable equipment to monitor astronauts' health and conduct research in microgravity. Instruments developed for space medicine, such as telemedicine systems, remote monitoring devices, and compact diagnostic tools, have been adapted for use on Earth. These innovations improve healthcare access in remote or underserved regions, enabling doctors to monitor patients, deliver care, and respond to emergencies efficiently. Space-derived medical devices enhance precision, portability, and reliability, improving patient outcomes worldwide. By translating technologies from space health research to Earth-based applications, lives are saved and healthcare becomes more equitable.

Energy solutions have also emerged from space-derived technologies. Solar panels designed for spacecraft have inspired high-efficiency photovoltaic cells used in renewable energy systems on Earth. Advanced thermal management systems, initially designed for satellites, optimize energy efficiency in buildings, vehicles, and industrial processes. Space technology drives innovation in energy storage, distribution, and sustainable design, allowing society to reduce reliance on fossil fuels. By applying engineering solutions from space missions to terrestrial energy systems, industries achieve greater efficiency and environmental sustainability. Space research accelerates the transition to clean, reliable energy worldwide.

Environmental monitoring and management are powered by space-derived tools. Satellites and remote sensing systems developed for Earth observation provide critical data for climate science, disaster prediction, water resource management, and pollution control. These technologies allow governments, NGOs, and private enterprises to track environmental changes in real time and make informed decisions. Space-based imaging and sensor networks enable proactive management of forests, oceans, and agricultural lands, helping to preserve ecosystems and biodiversity. By leveraging space-derived monitoring systems, society gains the ability to protect natural resources more effectively. Space research ensures that technological progress also supports environmental stewardship.

Advanced manufacturing technologies from space programs have transformed industrial production. Techniques developed to produce spacecraft components, such as additive manufacturing, precision machining, and lightweight material fabrication, are now used in automotive, aerospace, and consumer goods industries. These innovations reduce waste, improve product performance, and shorten production cycles. Industries applying space-derived manufacturing methods gain a competitive advantage while enhancing sustainability. Space exploration encourages the development of technologies that optimize efficiency, quality, and adaptability in production processes. The innovations born in space continue to drive industrial excellence on Earth.

Artificial intelligence and data analytics developed for space missions have wide-ranging terrestrial applications. AI systems created to process satellite imagery, control spacecraft, and predict mission outcomes now enhance logistics, agriculture, healthcare, and urban planning. These tools analyze large datasets, detect patterns, and provide actionable insights that improve decision-making. Space-derived AI technologies enable more precise forecasting, smarter resource allocation, and better risk management. By adapting intelligent systems from space missions, industries solve complex problems faster and more efficiently. Space research demonstrates how advanced analytics can transform society's ability to respond to challenges.

Space-derived innovations are fueling entrepreneurship and the creation of new industries. Technologies initially developed for spacecraft, satellites, and space missions often become the foundation for startups and commercial ventures on Earth. Companies specializing in satellite imagery, robotics, advanced materials, and AI analytics have emerged directly from space research. Entrepreneurs leverage these innovations to address challenges in healthcare, agriculture, energy, and environmental management. By transforming space technology into market-ready solutions, businesses create jobs, drive economic growth, and deliver societal benefits. Space-derived innovation stimulates both technological progress and economic opportunity, creating a dynamic cycle of discovery and application.

The commercialization of satellite technology has created new markets and services. Earth observation data, once reserved for scientific and military purposes, now supports agriculture, disaster response, urban planning, and climate research. Startups provide tailored analytics for precision farming, resource management, and environmental monitoring, enabling organizations to make informed, efficient decisions. Satellite-enabled connectivity has expanded access to remote areas, supporting education, commerce, and telemedicine. Space technology has unlocked entrepreneurial opportunities that address real-world problems while generating economic value. The global economy benefits from innovations that originated in the pursuit of exploration beyond Earth.

Spin-offs from space robotics are driving industrial automation and innovation. Robotic systems designed for satellite repair, space station operations, and planetary exploration are adapted for manufacturing, logistics, and healthcare on Earth. Automation solutions enhance productivity, precision, and safety in environments that are hazardous or labor-intensive. Startups and established companies alike adopt space-derived robotics to optimize processes, reduce costs, and improve efficiency. By commercializing technologies proven in extreme space conditions, industries gain access to highly reliable and innovative solutions. Space research catalyzes entrepreneurship by providing tools that accelerate industrial transformation.

Advanced materials developed for spacecraft inspire new products and applications in multiple sectors. Lightweight composites, heat-resistant alloys, and radiation-shielding materials improve automotive, aerospace, electronics, and medical device performance. Startups and research institutions license or adapt these materials to develop next-generation products that are stronger, more efficient, and more durable. Space-derived materials open opportunities for innovation in sustainable construction, high-performance manufacturing, and consumer goods. By translating materials science from space to Earth, entrepreneurs create products that enhance everyday life. Space exploration drives material innovation that powers technological and commercial advancement.

The global economy benefits from knowledge transfer from space research to Earth industries. Universities, research labs, and space agencies collaborate with private companies to convert space technologies into marketable products. This knowledge transfer strengthens innovation ecosystems, fosters technological competitiveness, and stimulates economic growth. Industries gain access to advanced techniques, tools, and expertise that would otherwise be unavailable. Space-derived technologies accelerate the pace of innovation, helping countries and companies remain globally competitive. By bridging exploration and commerce, space research creates value that extends far beyond orbit.

Space-derived technologies are accelerating solutions to critical societal challenges on Earth. In healthcare, innovations such as remote monitoring devices, imaging systems, and telemedicine platforms, originally developed for astronauts, now enhance patient care worldwide. Hospitals and clinics in remote or underserved regions benefit from space-inspired diagnostic and monitoring tools that improve treatment outcomes. In environmental protection, satellite monitoring and advanced sensors enable precise tracking of deforestation, pollution, and climate change. Governments and NGOs can implement evidence-based strategies to conserve ecosystems and mitigate environmental risks. Space-derived technologies transform how societies address health, environmental, and infrastructure challenges, creating tangible improvements in quality of life.

Infrastructure resilience is strengthened through space-derived innovation. Satellites and sensor networks monitor bridges, roads, dams, and energy systems, providing early warnings of structural weaknesses or environmental stress. Predictive analytics, originally designed for spacecraft and mission planning, now optimize maintenance, reduce downtime, and prevent disasters. By applying these technologies to urban planning and disaster preparedness, governments can build safer and more sustainable communities. Space-derived tools enable proactive management of infrastructure, enhancing both safety and operational efficiency. Space exploration provides the technical foundation for resilient, future-ready infrastructure systems.

In agriculture, space-based innovations improve productivity, sustainability, and food security. Satellite imagery, remote sensing, and AI analytics allow farmers to monitor soil health, crop growth, and water usage with unprecedented precision. Technologies initially developed for Earth observation in space programs enable predictive models that optimize planting schedules, irrigation, and resource allocation. By applying these innovations, agricultural systems reduce waste, increase yields, and minimize environmental impact. Space-derived tools transform farming into a data-driven, efficient, and sustainable practice. Space exploration thus directly supports global food security and responsible resource management.

Energy systems benefit from technologies originally developed for spacecraft and satellites. Solar power, battery storage, thermal management, and energy-efficient designs pioneered for space missions are now applied to renewable energy projects, grid optimization, and energy conservation. Advanced sensors and monitoring systems derived from satellites allow utilities to manage resources effectively and detect inefficiencies or faults quickly. These innovations increase energy reliability, reduce costs, and support the transition to sustainable power sources. Space technology accelerates energy innovation, ensuring more sustainable and resilient systems for society.

Disaster response is enhanced by space-derived technologies that integrate satellite imagery, real-time communication, and predictive analytics. Emergency management teams can assess damage, coordinate relief operations, and track recovery progress efficiently. Drones and robotics inspired by space exploration assist in search and rescue, supply delivery, and damage assessment. Predictive modeling helps anticipate secondary disasters, enabling faster and more effective interventions. By leveraging space-derived solutions, humanitarian and government agencies improve resilience, reduce risk, and save lives. Space technology bridges the gap between scientific innovation and immediate societal benefit.

Space-derived technology drives innovation, economic growth, and solutions to some of the world's most pressing challenges. By translating innovations from satellites, spacecraft, and space missions to terrestrial applications, industries gain advanced tools that improve efficiency, safety, and sustainability. Medical devices, robotics, energy systems, advanced materials, and data analytics—all inspired by space research—address critical needs in healthcare, infrastructure, agriculture, and environmental management. Startups and established companies alike leverage these technologies to create products and services that enhance human life and generate economic value. Space-derived innovation accelerates problem-solving and ensures that knowledge gained in exploration benefits society on Earth.

Entrepreneurship is fueled by technologies originally designed for the extreme conditions of space. Spin-offs and commercial ventures emerge from innovations in materials science, AI, robotics, and communications. Companies develop market-ready solutions that solve real-world problems, creating jobs and driving economic development. The process of converting space technology into commercial applications strengthens research ecosystems, promotes collaboration between academia, government, and industry, and fosters global competitiveness. Space exploration catalyzes economic growth by transforming scientific discovery into practical, impactful solutions.

Societal challenges are addressed more effectively through space-inspired solutions. Satellite monitoring, remote sensing, and predictive analytics provide tools for disaster response, environmental conservation, and urban planning. Space-derived innovations enhance resilience, efficiency, and sustainability, enabling governments and organizations to act proactively rather than reactively. Communities benefit from improved infrastructure, optimized resource management, and advanced medical and energy solutions. Space technology ensures that innovation directly translates into tangible improvements in quality of life.

Global collaboration in space research amplifies the impact of innovation. Countries, universities, and private companies share expertise, data, and technology, accelerating the pace of development. International partnerships allow technologies developed for one context to be adapted and applied globally, addressing challenges that transcend borders. Knowledge exchange fosters creativity, reduces duplication of effort, and ensures that innovations benefit humanity broadly. Space exploration serves as a unifying force for global scientific and technological progress.

Ultimately, space-derived technology exemplifies the power of exploration to create real-world benefits. The rigorous demands of space missions drive innovations that are efficient, resilient, and adaptable, transforming industries and society on Earth. By harnessing the knowledge, tools, and methods developed for space, humanity can solve complex problems, accelerate economic growth, and improve quality of life worldwide. Space exploration not only expands our understanding of the cosmos but also delivers solutions that make life on Earth safer, healthier, and more sustainable. Satellites, spacecraft, and spin-off technologies demonstrate that the pursuit of knowledge beyond our planet directly enriches life within it.

Chapter . 23

Medical Advancements Inspired by Space Research

Applying insights from space medicine to improve treatments and healthcare on Earth.

Medical research conducted in space has led to innovations that improve health and medical care on Earth. Astronauts experience unique physiological challenges in microgravity, radiation, and confined environments, prompting the development of advanced monitoring systems, diagnostic tools, and treatments. Scientists study bone density loss, muscle atrophy, cardiovascular changes, and immune system responses to understand the effects of extreme conditions on human health. These insights inform treatments for patients with similar conditions on Earth, such as osteoporosis, muscle degeneration, and immune disorders. By translating space research into terrestrial healthcare solutions, medical professionals gain tools to prevent, monitor, and treat diseases more effectively. Space medicine demonstrates the tangible benefits of applying knowledge gained beyond Earth to improve human health.

Telemedicine and remote monitoring technologies are key contributions from space medicine. Astronauts rely on real-time health monitoring systems that transmit data to medical teams on Earth. These technologies have been adapted for patients in remote, rural, or underserved areas, enabling doctors to monitor vital signs, detect abnormalities, and respond to emergencies without physical presence. Wearable sensors, compact diagnostic devices, and wireless monitoring tools have roots in space research. By extending medical expertise to patients regardless of location, space-inspired technologies increase access to healthcare and reduce response times. Space research has transformed the delivery of medical care by making it more precise, timely, and accessible.

Radiation protection innovations developed for space missions benefit medical treatments on Earth. Space environments expose astronauts to high-energy cosmic radiation, prompting the creation of shielding materials and protective strategies. These developments inform radiation therapy techniques used to treat cancer patients safely. Advances in dosimetry, protective gear, and treatment planning enhance the precision and safety of radiation-based medical procedures. Patients receive targeted therapy while minimizing damage to surrounding healthy tissue. Space-derived radiation solutions demonstrate how research in extreme conditions can improve medical safety and treatment outcomes on Earth.

Bone and muscle health research in space has directly influenced therapies for degenerative diseases. Microgravity causes astronauts to lose bone density and muscle mass at accelerated rates. Experiments in space have led to the development of countermeasures, including resistance exercise protocols, pharmaceuticals, and nutritional strategies. These findings inform treatments for osteoporosis, sarcopenia, and other musculoskeletal disorders. Patients on Earth benefit from improved preventive care, rehabilitation techniques, and physical therapy methods derived from space research. Space medicine provides critical insights into maintaining physical health in challenging conditions, which are directly applicable to Earth-bound medical care.

Advanced imaging and diagnostic tools developed for monitoring astronaut health have applications in terrestrial medicine. Non-invasive imaging systems, portable ultrasound devices, and miniaturized diagnostic instruments were designed to provide accurate assessments in confined and resource-limited space environments. Hospitals, clinics, and field medical teams now use these technologies to improve diagnostics, reduce invasive procedures, and increase patient safety. Space-driven innovations enhance the accuracy, portability, and efficiency of medical care, particularly in areas lacking extensive infrastructure. By translating space technologies to healthcare, patients receive faster, safer, and more effective treatments.

Case studies demonstrate how space-inspired medical technologies improve healthcare outcomes on Earth. One example is the use of telemedicine and remote monitoring systems, originally developed for astronauts, to provide care in rural or disaster-affected regions. Patients with chronic illnesses can be continuously monitored through wearable devices, allowing doctors to detect early signs of complications and intervene promptly. Hospitals use satellite-enabled communication networks to provide consultations and coordinate emergency responses in real time. These systems reduce travel burdens, increase access to specialists, and save lives. Space research has shown that health monitoring technology designed for extreme conditions can transform patient care across diverse settings.

Rehabilitation techniques for musculoskeletal disorders have benefited from research conducted in space. Astronauts in microgravity environments experience rapid bone density loss and muscle atrophy, leading scientists to develop specialized exercise regimens and mechanical resistance devices. These innovations are now applied to patients recovering from injury, surgery, or conditions like osteoporosis. Physical therapy programs use resistance-based machines and wearable monitoring devices to track progress and optimize recovery. By adapting strategies from space medicine, clinicians improve patient mobility, strength, and overall health outcomes. Space research informs rehabilitation methods that accelerate recovery and enhance quality of life.

Surgical innovations have been influenced by space-derived technologies. Miniaturized medical instruments, robotic-assisted surgery systems, and portable diagnostic tools were initially developed for use in spacecraft and confined environments. On Earth, these tools allow surgeons to perform precise procedures with minimal invasiveness, reducing complications and recovery times. Remote surgery applications, using high-speed communication networks inspired by space telemetry systems, enable experts to guide operations in locations lacking specialized surgeons. By transferring space technologies to the surgical field, healthcare becomes safer, more accurate, and accessible to underserved populations. Space research advances the frontiers of modern surgical care.

Cardiovascular health monitoring also benefits from space research. Microgravity affects fluid distribution and heart function in astronauts, leading to the development of advanced cardiac monitoring systems. These systems track heart rate, blood pressure, and vascular changes continuously. On Earth, similar technologies are applied to patients with heart disease, hypertension, and other circulatory conditions. Remote monitoring devices allow healthcare providers to detect early warning signs of cardiac events and implement timely interventions. Space-derived cardiovascular technologies enhance disease prevention, diagnosis, and treatment.

Immunology and infection control research in space has led to advancements in medical practices on Earth. Space environments alter immune system function, prompting the development of techniques to monitor immune responses and prevent infections. These insights inform treatments for immunocompromised patients, vaccine strategies, and hospital infection control protocols. By understanding how extreme environments affect immunity, researchers develop therapies that strengthen defense mechanisms and improve patient outcomes. Space medicine provides critical knowledge that enhances global healthcare systems.

Pharmaceutical research inspired by space medicine has led to new drug formulations and delivery methods. Microgravity affects how drugs dissolve, mix, and act on the human body, prompting scientists to develop stable, effective medications for space. These formulations have been adapted for Earth-based healthcare, improving the stability, bioavailability, and efficiency of critical drugs. Controlled-release systems and compact drug delivery devices, initially designed for space missions, now benefit patients with chronic conditions who require precise dosing. By applying lessons from space pharmacology, medical providers ensure that medications are safer, more effective, and easier to administer. Space research drives innovation in drug development and patient-centered therapies.

Nutrition research in space has also influenced health care on Earth. Astronauts require specialized diets to maintain bone density, muscle mass, and overall health in microgravity. Scientists developed nutrient-rich, easily digestible, and compact foods that meet these unique needs. These dietary strategies have been adapted for patients on restrictive diets, elderly populations, and individuals with malabsorption or chronic illnesses. Nutritional supplements, fortified foods, and optimized meal planning now help prevent deficiencies and support recovery. Space-inspired nutrition research demonstrates that optimizing diet for extreme conditions can enhance health and performance in everyday life.

Mental health and cognitive performance are critical areas addressed through space medicine research. Astronauts face isolation, confinement, and high-stress environments, leading to the development of psychological support systems, monitoring tools, and stress reduction interventions. Virtual reality simulations, cognitive training programs, and remote counseling technologies, designed for space crews, have been adapted for patients on Earth. These innovations improve mental resilience, reduce stress, and support recovery from trauma or illness. Space-inspired mental health solutions highlight the importance of holistic care, integrating mind and body for optimal well-being.

Portable diagnostic tools from space research are revolutionizing healthcare accessibility. Devices designed for spacecraft, capable of measuring vital signs, detecting pathogens, and analyzing samples in limited space, are now used in remote clinics, emergency response units, and low-resource settings. Point-of-care testing allows healthcare providers to make immediate, evidence-based decisions without relying on centralized laboratories. These technologies improve diagnostic speed, accuracy, and accessibility for populations worldwide. Space research demonstrates that innovation for extreme conditions can overcome terrestrial limitations in healthcare delivery.

Case studies highlight how space-derived medical innovations are transforming healthcare in hospitals, disaster zones, and remote areas. In rural clinics, telemedicine systems developed for astronauts allow doctors to monitor patients, provide consultations, and adjust treatments in real time. Patients with chronic illnesses, limited mobility, or rare conditions benefit from continuous monitoring without the need to travel long distances. Satellite communication ensures these systems remain operational even in areas with weak infrastructure. By leveraging space-inspired technology, healthcare providers can deliver timely, effective care to underserved populations. These systems demonstrate how innovations from space exploration directly improve patient outcomes on Earth.

In disaster zones, portable diagnostic tools and remote monitoring devices save lives. Following natural disasters, hospitals and emergency teams often face limited access to lab facilities and medical equipment. Devices designed for use in space, which are compact, reliable, and multifunctional, allow medical personnel to conduct rapid assessments and administer treatment efficiently. For example, portable ultrasound devices, blood analyzers, and vital-sign monitors provide critical information in real time. Space-derived innovations enhance disaster response, reduce treatment delays, and improve survival rates. They demonstrate that technologies developed for extreme conditions can be adapted for high-stakes, unpredictable environments on Earth.

Urban hospitals have also benefited from space-inspired medical technologies. Robotic-assisted surgery systems, advanced imaging tools, and minimally invasive diagnostic devices were initially developed for astronaut care in constrained environments. These innovations enable surgeons to perform complex procedures with higher precision, reduced risk, and shorter recovery times for patients. Hospitals have integrated space-derived monitoring systems to track patient vitals continuously, optimize treatment protocols, and improve workflow efficiency. By adopting space-inspired solutions, healthcare institutions increase patient safety, operational efficiency, and treatment effectiveness. Space research provides practical tools that revolutionize hospital care.

Telehealth platforms influenced by space medicine are improving care in developing regions. Communities with limited access to specialists can connect with urban hospitals and global experts through satellite communication networks. Remote monitoring, data transmission, and virtual consultations enable patients to receive expert guidance without leaving their local clinics. These systems are particularly valuable for managing chronic diseases, prenatal care, and emergency response. Space-derived communication and monitoring technologies ensure that high-quality healthcare is available to populations previously underserved. Satellites and space-inspired devices make equitable medical care possible worldwide.

The impact of space medicine on healthcare innovation and patient outcomes is profound and far-reaching. Insights gained from studying human health in microgravity and extreme environments have led to the development of advanced diagnostic tools, telemedicine systems, wearable sensors, and rehabilitation devices. These innovations enhance patient care, increase accessibility to medical services, and improve treatment outcomes worldwide. Hospitals, rural clinics, disaster zones, and remote communities now benefit from technologies originally designed to keep astronauts healthy in space. By translating space research into practical applications, healthcare providers can deliver faster, more precise, and more effective interventions. Space medicine demonstrates that research beyond Earth has immediate, tangible benefits for life on our planet.

Medical advancements inspired by space research also contribute to preventive care. Continuous monitoring systems, advanced imaging, and wearable sensors allow early detection of diseases and health anomalies. By identifying issues before they escalate, these technologies reduce hospitalizations, lower healthcare costs, and improve long-term outcomes. Patients receive personalized, data-driven care informed by space-derived technologies. Space research emphasizes proactive, rather than reactive, healthcare strategies that enhance quality of life. The principles learned from astronaut health management now guide better health practices for populations worldwide.

Global health accessibility is further improved through space-inspired innovations. Telemedicine, satellite-enabled communication, and portable diagnostic devices ensure that patients in remote or underserved regions receive timely medical attention. Space technologies bridge the gap between medical expertise and geographic limitations. Populations previously excluded from specialized healthcare can now access diagnostics, consultations, and treatment guidance. By leveraging technologies developed for extreme space conditions, healthcare systems provide equitable, life-saving services to communities across the globe. Space medicine exemplifies the potential of technology to make healthcare universally accessible.

Space medicine also drives interdisciplinary collaboration in healthcare innovation. Engineers, biologists, physicians, and data scientists work together to develop tools that improve monitoring, diagnostics, and treatment outcomes. These collaborations generate solutions that are robust, scalable, and adaptable to diverse medical environments. By fostering cooperation across disciplines, space medicine accelerates the development of practical technologies for real-world healthcare challenges. The collaborative spirit of space research ensures that medical innovation is both rapid and impactful.

Ultimately, medical advancements inspired by space research demonstrate how exploration beyond Earth can directly enhance human health on our planet. From diagnostic tools and telemedicine systems to rehabilitation devices and preventive care strategies, space medicine provides technologies that save lives, improve recovery, and extend access to healthcare. Satellites, spacecraft, and space research initiatives have created a foundation for medical innovation that benefits everyone, everywhere. By applying lessons learned from astronauts and space environments, humanity can advance global health, optimize patient care, and ensure that the knowledge gained beyond our planet translates into tangible, life-saving solutions on Earth.

Chapter . 24

Improving Materials and Manufacturing Through Space Technology

Using techniques developed for spacecraft to create stronger, more sustainable products on Earth.

Space technology has revolutionized materials science and manufacturing processes, leading to innovations that enhance durability, efficiency, and sustainability on Earth. The extreme conditions of space—such as microgravity, radiation, and temperature extremes—require materials that are lightweight, strong, and resistant to wear. Engineers and scientists have developed specialized alloys, composites, and polymers for spacecraft and satellite construction. These materials often exceed the performance of traditional Earth-based alternatives. By adapting these innovations, industries can produce products that are longer-lasting, more resilient, and environmentally responsible. Space technology thus drives the evolution of manufacturing from conventional methods to advanced, high-performance solutions.

Additive manufacturing, or 3D printing, is one of the most transformative technologies derived from space research. NASA and other space agencies developed 3D printing systems capable of producing components in microgravity for satellites, spacecraft, and space station modules. These techniques allow precise, customizable fabrication using minimal materials. On Earth, industries utilize space-derived additive manufacturing for aerospace, automotive, and medical device production. 3D printing reduces waste, lowers production costs, and enables the creation of complex structures that would be impossible with traditional methods. Space-driven 3D printing demonstrates how manufacturing in extreme environments can inspire efficiency and innovation on Earth.

Advanced composites and lightweight materials developed for spacecraft are now used across multiple industries. Carbon-fiber reinforced polymers, high-strength alloys, and thermal-resistant coatings were initially designed to withstand the rigors of launch, re-entry, and long-term exposure to space. These materials enhance the performance of vehicles, construction projects, sporting equipment, and electronic devices on Earth. Their combination of strength, durability, and weight reduction improves energy efficiency and sustainability in manufacturing. Space research illustrates how solving extreme engineering challenges can yield materials that transform industries worldwide.

Precision manufacturing techniques from space programs have also been adapted for Earth-based industries. Spacecraft components require micron-level accuracy and tolerance to ensure proper function in orbit. These rigorous standards drove innovations in machining, surface finishing, and assembly processes. Industries now apply these techniques to produce medical instruments, electronics, and high-performance machinery with exceptional quality. By adopting space-grade manufacturing methods, companies achieve reliability, efficiency, and superior product performance. Space technology raises the standard for precision and consistency in industrial production.

Environmental sustainability is a key benefit of space-inspired materials and manufacturing. Space missions require minimal waste and efficient resource utilization, prompting the development of processes that conserve energy and materials. Recycling techniques, modular design, and lightweight construction are directly transferable to Earth-based industries. By reducing material consumption and extending product lifespans, these innovations lower environmental impact while maintaining performance. Space technology provides a blueprint for sustainable manufacturing practices that meet modern economic and ecological demands.

Case studies demonstrate how space-derived materials and manufacturing techniques are transforming industries on Earth. In aerospace, carbon-fiber composites and heat-resistant alloys originally developed for spacecraft are now standard in commercial aircraft, drones, and satellites. These materials reduce weight, improve fuel efficiency, and increase durability, enabling safer and more cost-effective air travel. Aerospace manufacturers also adopt additive manufacturing techniques from space programs to produce complex parts with fewer components and minimal waste. By applying innovations developed for space, the aerospace industry benefits from enhanced performance and sustainability. Space research has thus accelerated advancements that make aviation safer, more efficient, and environmentally responsible.

The automotive industry has also embraced space-inspired materials and manufacturing. Lightweight alloys, high-strength polymers, and precision-engineered components derived from spacecraft design reduce vehicle weight and increase fuel efficiency. Electric vehicle manufacturers use space-grade thermal management materials to optimize battery performance and longevity. Additive manufacturing techniques allow rapid prototyping and the production of complex, high-performance parts. These innovations improve safety, efficiency, and sustainability while reducing production costs. Space technology demonstrates how solutions designed for extreme conditions can drive innovation in everyday transportation systems.

Medical device manufacturing benefits from precision and materials innovations originating in space programs. Lightweight, biocompatible materials developed for spacecraft and satellites are now used in surgical instruments, prosthetics, and implants. Additive manufacturing enables the creation of patient-specific devices that improve comfort and functionality. Space-derived techniques ensure high precision, reliability, and safety in medical applications. Hospitals and clinics benefit from tools and devices that are both advanced and durable. Space-inspired manufacturing enhances healthcare by providing innovative, high-quality solutions for patient care.

Consumer products have also been improved through space-inspired technologies. Sports equipment, electronics, and household tools now incorporate advanced composites, lightweight alloys, and thermal-resistant materials originally developed for space missions. Products are more durable, energy-efficient, and functional, enhancing user experience while reducing environmental impact. Additive manufacturing allows for customization and rapid prototyping, meeting consumer demand for personalized, high-quality products. Space technology thus influences everyday life, providing benefits far beyond the laboratory or industrial facility.

Industrial applications, such as energy and construction, leverage space-derived materials and processes. Thermal-resistant coatings and lightweight structural components improve efficiency in power generation, machinery, and buildings. Precision manufacturing techniques ensure consistent quality, reducing material waste and maintenance costs. Industries adopting space-inspired solutions achieve improved performance, reliability, and sustainability. By translating the demands of space missions to Earth-based production, industries can innovate more effectively and responsibly. Space research creates technologies that optimize both industrial output and environmental stewardship.

Space technology has also advanced the development of nanomaterials with unique properties. Materials engineered at the nanoscale for spacecraft, satellites, and sensors exhibit exceptional strength, conductivity, and thermal resistance. These nanomaterials are now applied in electronics, medicine, energy storage, and aerospace industries on Earth. For example, nanoscale coatings improve the durability of solar panels and electronic devices, while nanoparticle-based materials enhance battery efficiency and lifespan. By studying materials in space, scientists have discovered ways to manipulate structures at the molecular level for superior performance. Space research demonstrates that innovations in extreme conditions can inspire breakthroughs in microscopic engineering for Earth-based applications.

Robotics-assisted manufacturing, originally designed for space missions, has transformed industrial production on Earth. Robotic arms, automated assembly systems, and remote-controlled tools were developed for constructing satellites, repairing spacecraft, and operating in environments with limited human intervention. These technologies are now applied to automotive assembly lines, electronics production, and precision machining. Robotics increase efficiency, reduce human error, and allow the creation of highly complex components that would be difficult or impossible to produce manually. Space-inspired automation ensures consistent quality and accelerates production timelines. Industries adopting robotics derived from space programs achieve both precision and scalability in manufacturing.

Automation systems developed for space applications enhance safety and productivity in Earth-based industries. Satellite monitoring, sensors, and AI-guided control systems, initially designed for spacecraft operations, allow factories and energy facilities to monitor processes in real time. Automated alerts and predictive maintenance prevent failures, reduce downtime, and optimize resource utilization. These systems improve operational efficiency, minimize waste, and ensure high-quality outcomes in sectors ranging from manufacturing to utilities. By integrating space-derived automation, industries achieve a level of reliability and optimization previously unattainable. Space technology provides a model for intelligent, adaptive systems that transform industrial efficiency.

Nanotechnology and robotics are increasingly integrated into sustainable manufacturing solutions. Lightweight composites, recyclable materials, and additive manufacturing techniques reduce energy use and material waste. Robotics and automation ensure that these processes are precise and efficient, minimizing environmental impact. Industries adopting space-inspired technologies are able to meet growing demands for sustainability, performance, and quality. Space research drives innovations that balance productivity with ecological responsibility. By leveraging materials and methods proven in space, manufacturers can achieve a sustainable future while meeting modern industrial needs.

Space-derived manufacturing techniques have greatly influenced energy production and sustainability efforts on Earth. Materials and processes developed for spacecraft and satellites enable the creation of more efficient solar panels, wind turbines, and energy storage systems. Lightweight, durable materials reduce energy loss and increase the lifespan of renewable energy infrastructure. Precision manufacturing techniques ensure that components operate at optimal efficiency while minimizing waste during production. Robotics-assisted assembly allows for the construction of complex energy systems with high accuracy and consistency. Space technology provides a blueprint for energy solutions that are efficient, reliable, and environmentally sustainable.

Construction and infrastructure industries also benefit from innovations inspired by space research. Advanced composites, lightweight alloys, and thermal-resistant coatings, initially designed for spacecraft, improve building materials and structural components. These materials reduce weight while increasing strength, durability, and resistance to environmental stressors. Additive manufacturing and modular assembly techniques enable faster construction, lower labor costs, and enhanced quality control. Space-derived precision and materials innovation ensure that infrastructure projects are safer, more resilient, and sustainable. The influence of space technology transforms construction practices, making them more efficient and environmentally responsible.

Sustainable infrastructure development relies on the principles of space technology. Engineers apply lessons from resource-constrained environments in space to optimize material usage and energy consumption in buildings and industrial projects. Advanced sensors and monitoring systems, adapted from satellites and spacecraft, track structural integrity, detect wear, and predict maintenance needs. These innovations improve safety, reduce operational costs, and extend the lifespan of infrastructure. By leveraging space-derived techniques, urban planners and engineers can create sustainable, resilient, and adaptable infrastructure systems. Space research provides practical solutions for meeting the challenges of modern urban development.

Space-based technologies also improve water and resource management within infrastructure projects. Materials developed for spacecraft, such as corrosion-resistant coatings and lightweight composites, are applied to water treatment systems, pipelines, and storage facilities. These innovations enhance durability, reduce leakage, and improve overall efficiency. Remote monitoring and predictive maintenance, inspired by space mission control systems, allow operators to optimize resource usage and respond to issues before failures occur. Space technology ensures that infrastructure systems are not only high-performing but also sustainable and environmentally responsible.

The overall impact of space-derived materials and manufacturing on industry, sustainability, and technological innovation is profound. Materials originally developed for spacecraft—lightweight composites, heat-resistant alloys, and advanced polymers—have been adapted for aerospace, automotive, medical, construction, and consumer industries. These innovations improve product performance, extend lifespans, and reduce resource consumption. Additive manufacturing, robotics-assisted assembly, and automation techniques enable precise, efficient, and sustainable production processes. By applying technologies designed for extreme conditions in space, industries achieve higher quality, lower costs, and reduced environmental impact. Space technology demonstrates that solving challenges in orbit can yield transformative solutions on Earth.

Sustainable manufacturing benefits from lessons learned in space research. Resource efficiency, waste reduction, and energy optimization are central to both spacecraft production and Earth-based manufacturing. Techniques developed for recycling, modular construction, and precision engineering allow industries to minimize environmental impact while maintaining high performance. Space-derived practices ensure that industrial innovation aligns with ecological responsibility. By leveraging space-inspired methods, companies can produce durable, high-quality goods without compromising sustainability. Space technology serves as a model for integrating efficiency and environmental stewardship in manufacturing.

Industrial competitiveness is strengthened through the adoption of space-inspired materials and processes. Companies that implement advanced composites, additive manufacturing, and robotics gain a technological edge over competitors. Products are lighter, stronger, and more efficient, while production is faster and more precise. The ability to innovate rapidly and sustainably allows industries to respond to evolving markets and global demands. Space research accelerates technological progress, ensuring that Earth-based industries remain at the forefront of innovation. The influence of space-derived manufacturing is both practical and strategic, driving economic growth and technological leadership.

Innovation ecosystems benefit from collaboration between space agencies, universities, and industry. Partnerships foster the transfer of knowledge, materials, and techniques developed for space applications to commercial and industrial settings. Startups, research labs, and manufacturing companies can experiment with space-derived technologies to create new products, optimize processes, and improve efficiency. This collaborative approach accelerates the pace of technological advancement and ensures that innovations reach a broad range of industries. Space exploration serves as a hub for innovation, connecting science, engineering, and commerce in ways that benefit society.

Ultimately, materials and manufacturing technologies derived from space research exemplify how exploration beyond Earth can transform life on our planet. From stronger and lighter materials to precise, automated production techniques, these innovations improve products, processes, and infrastructure across multiple sectors. Sustainability, efficiency, and resilience are hallmarks of space-inspired manufacturing, ensuring that industries operate responsibly and effectively. Space technology continues to provide the knowledge and tools necessary to advance industrial capability while addressing the challenges of a growing, resource-conscious world. By applying lessons from space, humanity can innovate, build, and sustain a better future on Earth.

Chapter . 25

Artificial Intelligence and Robotics for Earthly Challenges

Adapting space robotics and AI to automate tasks and enhance productivity on Earth.

Artificial intelligence (AI) and robotics, originally developed for space missions, are now transforming industries and daily life on Earth. Space agencies created autonomous systems to operate spacecraft, manage satellites, and assist astronauts in environments where human intervention is limited. These AI systems and robotic platforms were designed to make decisions, respond to unexpected conditions, and perform complex tasks in real time. On Earth, these innovations are applied in manufacturing, healthcare, agriculture, disaster response, and transportation. By adapting space-derived AI and robotics, societies can increase efficiency, reduce human risk, and address complex challenges more effectively. Space research provides a model for intelligent, autonomous systems that enhance productivity in diverse sectors.

Robotics developed for space missions are widely used in industrial automation. Robotic arms, mobile platforms, and autonomous assembly systems initially created for spacecraft construction and satellite servicing are now employed in factories and warehouses. These robots can handle repetitive, hazardous, or precision-based tasks, improving safety and productivity. AI-guided robotics allow real-time monitoring, adaptive decision-making, and fault detection in industrial processes. By leveraging these technologies, industries reduce human error, minimize production downtime, and optimize resource use. Space-inspired robotics demonstrate how advanced automation can revolutionize modern manufacturing.

Artificial intelligence from space applications enhances data analysis and predictive capabilities. Satellites and spacecraft generate massive amounts of data that require sophisticated AI algorithms for processing and interpretation. These systems can detect anomalies, forecast outcomes, and support decision-making in dynamic environments. On Earth, AI systems derived from space technology analyze financial markets, monitor environmental changes, optimize logistics, and improve healthcare diagnostics. By applying AI techniques proven in space, organizations can process complex datasets faster and make more informed, accurate decisions. Space research accelerates the adoption of AI for real-world problem-solving.

Healthcare is a sector particularly influenced by space-inspired AI and robotics. Telemedicine platforms, remote surgery systems, and robotic-assisted procedures were initially developed to support astronauts in remote or microgravity conditions. These innovations now allow doctors to perform precise surgeries, monitor patient health remotely, and deliver care in underserved areas. AI algorithms analyze patient data, detect patterns, and recommend treatment options based on real-time information. Space-derived AI and robotics enhance medical efficiency, safety, and accessibility, improving outcomes for patients worldwide. Space research provides both the technology and the methodology to advance healthcare on Earth.

Agriculture also benefits from AI and robotics adapted from space research. Autonomous drones, robotic harvesters, and AI-driven monitoring systems were originally designed to collect data and perform tasks in remote or extreme space environments. These technologies are now used to monitor soil health, track crop growth, optimize irrigation, and apply fertilizers or pesticides precisely. AI algorithms help farmers predict crop diseases, estimate yields, and plan harvesting schedules. By using space-inspired AI and robotics, agriculture becomes more efficient, sustainable, and resilient to environmental challenges. Space technology transforms traditional farming into a data-driven, automated system capable of meeting global food demands.

Space-derived AI and robotics are transforming disaster response and emergency management on Earth. Autonomous drones and robotic systems, initially developed for planetary exploration and satellite maintenance, are now deployed to assess damage, deliver supplies, and conduct search-and-rescue operations. AI algorithms analyze satellite imagery and sensor data to predict disaster patterns, identify high-risk areas, and optimize resource allocation. These technologies allow emergency teams to respond more quickly, coordinate effectively, and minimize human risk. By adapting space-based innovations, communities can recover faster and save lives during natural or man-made disasters. Space research demonstrates how autonomous systems can address critical challenges in unpredictable environments.

Environmental monitoring benefits from AI and robotics originally designed for space missions. Satellites and orbital sensors generate massive amounts of data on climate, deforestation, ocean currents, and pollution. AI algorithms process this data to identify trends, predict environmental risks, and guide conservation efforts. Robotic platforms, including drones and autonomous vehicles, collect ground-level data in remote or hazardous regions. By leveraging space-inspired AI and robotics, scientists and policymakers gain precise, actionable insights to protect ecosystems and manage natural resources sustainably. Space technology enables proactive environmental management, enhancing our ability to respond to global challenges.

Logistics and supply chain management are also improved through space-derived AI and robotics. Autonomous vehicles, routing algorithms, and robotic warehouse systems were initially developed to navigate complex space environments and manage satellite operations. These technologies now optimize transportation, inventory management, and delivery systems on Earth. AI predicts demand, adjusts routes in real time, and reduces fuel consumption, while robotics automate repetitive tasks in warehouses and production facilities. By applying space-based innovations, businesses improve efficiency, reduce operational costs, and minimize environmental impact. Space technology provides a framework for smarter, faster, and more sustainable logistics.

Urban management and smart city initiatives leverage AI and robotics inspired by space research. Autonomous monitoring systems, sensor networks, and predictive analytics, originally designed for spacecraft and satellites, are now used to optimize traffic flow, energy usage, and public safety. AI algorithms process data from thousands of sensors to detect anomalies, manage infrastructure, and respond to emergencies efficiently. Robotic systems assist with maintenance, inspections, and environmental monitoring in urban settings. By translating space-derived technologies to city management, communities achieve higher efficiency, safety, and sustainability. Space research demonstrates that intelligent systems can improve urban life on a large scale.

Education and research benefit from space-inspired AI and robotics as well. Universities and training centers use autonomous systems and AI-driven simulations to teach robotics, programming, and data analysis. Students gain hands-on experience with technologies originally developed for space missions. These educational applications cultivate the next generation of engineers, scientists, and innovators capable of applying space technologies to solve Earthly challenges. By integrating space-derived tools into education, society ensures continued innovation, knowledge transfer, and technological growth. Space research fosters both practical applications and the skills necessary to implement them effectively.

Industrial automation has been transformed by AI and robotics originally developed for space missions. Robotic arms, autonomous assembly systems, and precision-controlled machines were designed to operate in spacecraft assembly, satellite maintenance, and extreme environments. On Earth, these technologies automate complex production processes, improve precision, and reduce human error. AI-driven monitoring systems optimize workflows, detect defects in real time, and predict maintenance needs. Industries such as automotive, electronics, and aerospace benefit from increased efficiency, cost savings, and higher-quality outputs. Space-inspired automation demonstrates how robotics and AI can revolutionize industrial operations while maintaining safety and consistency.

Healthcare optimization has also been influenced by space-derived AI and robotics. Robotic-assisted surgery, telemedicine platforms, and AI-driven diagnostics were initially developed to monitor and support astronauts in isolated, high-risk environments. These systems now enhance surgical precision, enable remote patient care, and improve diagnostic accuracy on Earth. AI algorithms process patient data to detect abnormalities, recommend interventions, and predict treatment outcomes. Robotics streamline repetitive or labor-intensive tasks, allowing healthcare professionals to focus on complex decision-making. By applying space-inspired technologies, hospitals and clinics increase patient safety, efficiency, and accessibility. Space research provides models for integrating autonomous systems into healthcare delivery.

Precision agriculture is another area benefiting from space-inspired AI and robotics. Autonomous drones, robotic harvesters, and AI-driven monitoring systems were originally created to gather data and perform tasks in remote or harsh environments. These systems now monitor soil health, track crop growth, optimize irrigation, and detect pests or disease. AI algorithms predict crop yields, plan harvesting schedules, and guide resource allocation. Robotics automate labor-intensive agricultural processes, increasing productivity and reducing environmental impact. Space-derived technologies transform farming into a data-driven, efficient, and sustainable practice capable of meeting global food demands.

Energy systems leverage AI and robotics from space applications to optimize production, distribution, and consumption. Autonomous monitoring systems, predictive maintenance algorithms, and robotic inspection platforms were initially developed for spacecraft and satellites. On Earth, these innovations are applied to renewable energy facilities, power grids, and industrial energy systems. AI analyzes performance data to enhance efficiency, detect faults, and reduce waste, while robotic systems perform inspections and maintenance in hazardous or hard-to-reach areas. Space-inspired technologies increase energy reliability, safety, and sustainability. AI and robotics provide tools for managing complex energy networks more effectively.

Transportation and logistics have been revolutionized by space-derived AI and robotics. Autonomous navigation systems, route optimization algorithms, and fleet management software were initially designed to control spacecraft and coordinate satellite constellations. These systems now improve delivery networks, optimize shipping routes, and automate warehouse operations on Earth. AI predicts demand, identifies bottlenecks, and reduces fuel consumption, while robotics handle inventory, packaging, and loading tasks. Space-inspired technologies enhance the efficiency, reliability, and sustainability of transportation and logistics networks. Space research demonstrates how intelligent systems can solve complex operational challenges in real-world settings.

Environmental monitoring has been greatly enhanced by AI and robotics derived from space technology. Satellites, sensors, and autonomous data-gathering robots originally developed for space missions now collect detailed information on climate patterns, deforestation, ocean health, and pollution. AI algorithms process this data to identify trends, predict environmental risks, and support conservation strategies. Autonomous drones and robotic vehicles gather on-the-ground measurements in remote or hazardous areas. By applying space-inspired technologies, scientists can monitor ecosystems more effectively, respond to environmental threats quickly, and make informed policy decisions. Space research provides tools that enable proactive and precise environmental stewardship on Earth.

Disaster preparedness and response benefit from AI and robotics initially designed for space operations. Robotic platforms developed for spacecraft maintenance and planetary exploration are now used in search-and-rescue missions, damage assessment, and supply delivery during natural disasters. AI systems analyze satellite imagery and sensor data to predict disaster impacts, optimize resource allocation, and guide emergency response teams. By reducing human exposure to danger and improving coordination, these technologies save lives and mitigate economic losses. Space-derived innovations enhance resilience by allowing rapid, data-driven action in unpredictable, high-risk scenarios. Autonomous systems and AI provide critical support in protecting communities from emergencies.

Smart cities utilize AI and robotics inspired by space research to optimize urban management. Autonomous monitoring systems, sensor networks, and predictive algorithms, originally designed for spacecraft and satellites, are used to regulate traffic, manage energy consumption, and improve public safety. Robotic maintenance systems and AI-driven analytics ensure efficient operation of infrastructure, reduce waste, and enhance city resilience. Data-driven decision-making allows municipalities to anticipate issues, optimize resource allocation, and improve the quality of life for residents. Space-derived technologies demonstrate how intelligent systems can make urban environments safer, more efficient, and sustainable.

Global infrastructure management is enhanced through space-inspired AI and robotics. Large-scale projects, such as power grids, water distribution networks, and transportation systems, benefit from autonomous monitoring, predictive maintenance, and data analytics. AI algorithms detect potential failures, optimize resource use, and coordinate system operations. Robotic inspection platforms reduce human risk and improve maintenance efficiency. By adapting space-derived technologies, industries and governments ensure that critical infrastructure remains reliable, efficient, and sustainable. Space research provides a framework for managing complex systems on a global scale.

Education and workforce training also leverage AI and robotics from space programs. Autonomous systems, AI-driven simulations, and robotic platforms originally developed for space exploration are used to train engineers, scientists, and technicians on Earth. Students gain hands-on experience with advanced technologies, preparing them to implement AI and robotics in diverse industries. By integrating space-inspired tools into education and professional development, societies cultivate the next generation of innovators capable of applying advanced technologies to solve Earthly challenges. Space research not only produces technologies but also fosters expertise and human capacity for innovation.

The overall impact of space-derived AI and robotics on Earthly industries, public services, and global problem-solving is profound. Technologies initially developed to operate spacecraft, satellites, and robotic explorers are now applied across manufacturing, healthcare, agriculture, energy, and disaster management. Autonomous systems reduce human risk, optimize processes, and improve efficiency, while AI algorithms provide predictive insights that guide decision-making. By integrating robotics and AI, industries can increase productivity, reduce errors, and create safer work environments. Space-inspired technologies demonstrate that innovations developed for extreme environments can have transformative applications on Earth.

Healthcare continues to benefit from AI and robotics adapted from space research. Robotic-assisted surgery, remote monitoring, and AI-driven diagnostics improve treatment precision, accessibility, and patient outcomes. Telemedicine platforms initially designed for astronauts allow doctors to care for patients in remote and underserved regions. AI analyzes medical data in real time, enabling early detection of disease and personalized treatment plans. Space-derived technologies improve the safety, efficiency, and equity of healthcare systems worldwide. Space research illustrates how intelligent systems can enhance human health in practical, scalable ways.

Disaster response and environmental management are strengthened by space-inspired AI and robotics. Autonomous drones, robotic platforms, and satellite-based AI systems collect critical data, predict risks, and guide interventions in emergencies. By reducing human exposure to hazards and enabling faster, data-driven responses, these technologies save lives and protect resources. Environmental monitoring, pollution control, and resource management all benefit from real-time data analysis and automation. Space-derived innovations allow societies to anticipate and mitigate risks, creating more resilient and sustainable communities.

Urban management, logistics, and infrastructure operations are also revolutionized by space-based AI and robotics. Intelligent monitoring systems, predictive maintenance, and autonomous robotics optimize traffic flow, energy usage, water distribution, and supply chains. Cities and industries can operate more efficiently, reduce waste, and respond proactively to challenges. Space-inspired technologies enable the management of complex systems at local, regional, and global scales. By applying lessons from space, Earth-bound infrastructure becomes smarter, safer, and more sustainable.

Ultimately, AI and robotics derived from space research exemplify the practical benefits of space exploration for life on Earth. From industrial automation and healthcare optimization to disaster preparedness and environmental stewardship, these technologies provide solutions to some of humanity's most pressing challenges. Space research not only advances technology but also inspires new methods of problem-solving, collaboration, and innovation. By harnessing the power of space-derived AI and robotics, societies can address complex challenges, improve quality of life, and create a safer, more sustainable future for all. The exploration of space proves that intelligent, autonomous systems designed for extreme conditions can deliver transformative benefits here on Earth.

Part VI: Sustainable Development and Global Impact

Chapter . 26

Supporting Climate Adaptation with Space Technology

Using orbital data to inform strategies that help communities adapt to climate change.

Space technology plays a crucial role in helping humanity adapt to the impacts of climate change. Satellites orbiting Earth collect vast amounts of data on temperature, precipitation, ocean currents, ice cover, and atmospheric conditions. This information allows scientists and policymakers to monitor environmental changes in real time, identify trends, and make informed decisions. Communities, governments, and organizations can use this data to implement strategies that reduce vulnerability and increase resilience. By leveraging space-derived insights, societies can anticipate climate risks, plan adaptation measures, and allocate resources effectively. Space technology ensures that climate adaptation efforts are grounded in accurate, comprehensive information.

One of the most important applications of space technology is in monitoring sea-level rise. Satellites equipped with radar altimeters provide precise measurements of ocean heights across the globe. This data allows coastal planners to assess flood risks, evaluate potential inundation zones, and design protective infrastructure such as seawalls, levees, and natural buffers. Early warning systems informed by satellite data enable communities to prepare for storm surges and extreme weather events. By integrating orbital information into planning, decision-makers can develop strategies that reduce economic and human losses. Space technology empowers societies to respond proactively to the threats posed by rising seas.

Agricultural adaptation also benefits from space technology. Satellites provide data on soil moisture, rainfall patterns, temperature, and vegetation health, which farmers use to optimize crop selection, planting schedules, and irrigation practices. AI algorithms analyze satellite imagery to detect early signs of drought, disease, or pest infestations. This information allows farmers to implement interventions that maintain productivity and reduce crop losses. By using space-derived data, agricultural systems can become more resilient to climate variability, ensuring food security even under changing conditions. Space technology enhances both planning and operational decision-making in farming.

Water resource management relies heavily on satellite data for climate adaptation. Spaceborne sensors track river flows, groundwater levels, and reservoir capacities, enabling efficient allocation and conservation of freshwater resources. Hydrologists use this information to predict droughts, floods, and seasonal water availability, guiding policy and infrastructure development. Urban and agricultural planners can optimize water distribution and storage, reducing shortages and minimizing environmental damage. Space technology provides the real-time monitoring and predictive tools needed to manage a resource that is increasingly stressed by climate change. Communities can therefore adapt more effectively to shifts in water availability.

Urban planning and infrastructure development are enhanced through space-based climate insights. Satellites provide data on temperature variations, urban heat islands, precipitation patterns, and storm trajectories. City planners can use this information to design resilient buildings, transport networks, and green spaces. Predictive models derived from satellite observations guide zoning decisions, emergency preparedness, and energy management. By integrating space-derived data into urban development, cities can reduce vulnerability to extreme weather and improve overall sustainability. Space technology enables smarter, more adaptive infrastructure planning in the face of climate challenges.

Case studies illustrate how space technology supports climate adaptation across diverse regions and communities. In Bangladesh, satellite data is used to monitor river flows, rainfall patterns, and storm surges, providing early warning systems for floods and cyclones. Communities receive alerts through mobile networks, enabling timely evacuation and preparation. Agricultural planners use the same data to adjust planting schedules and crop selection, reducing losses from extreme weather events. By integrating satellite observations into national planning, Bangladesh improves resilience to climate-related disasters. Space technology provides actionable intelligence that saves lives and protects livelihoods.

In Africa, space-based monitoring informs water management and agricultural adaptation. Countries like Kenya, Ethiopia, and Nigeria rely on satellites to track rainfall variability, soil moisture, and drought conditions. Farmers receive guidance on irrigation, crop selection, and planting timing, reducing the impact of climate shocks. Governments use this data to allocate resources, plan emergency responses, and develop long-term water strategies. AI algorithms analyze satellite imagery to forecast potential crop failures and food insecurity risks. By leveraging space technology, African nations enhance resilience in agriculture-dependent communities. Satellite data enables informed decision-making that sustains food security under changing climate conditions.

Coastal cities such as Miami, Jakarta, and Amsterdam use space technology to address sea-level rise and storm risks. Satellites provide high-resolution data on tidal patterns, coastal erosion, and storm surge potential. City planners integrate these observations into the design of seawalls, floodgates, and wetland restoration projects. Real-time satellite monitoring allows authorities to respond proactively to extreme weather, minimizing damage and loss of life. Predictive modeling derived from orbital data guides investment in resilient infrastructure. By using space technology, urban centers can adapt to rising seas and increasingly severe storms. These insights support sustainable development and protect vulnerable populations.

Indigenous communities also benefit from space-derived climate adaptation strategies. In Arctic regions, satellites monitor ice thickness, snow cover, and permafrost changes, providing critical information for transportation, hunting, and settlement planning. Communities can anticipate hazards such as ice collapse or thawing permafrost, adjusting traditional practices and infrastructure use. Data is often shared in accessible formats, combining scientific observations with traditional knowledge. By integrating satellite technology into local decision-making, indigenous populations increase safety and preserve cultural practices. Space technology empowers communities to adapt effectively while respecting local traditions.

Forest management and wildfire prevention are enhanced through space-based observations. Satellites track vegetation health, soil moisture, and fire-prone conditions across vast forested areas. AI-driven analysis predicts potential fire outbreaks, enabling preemptive interventions such as controlled burns or resource allocation for firefighting. Governments and conservation organizations use these insights to protect ecosystems, mitigate carbon emissions, and safeguard communities. Space technology allows for continuous, large-scale monitoring that would be impossible through ground-based methods alone. By providing early warnings and strategic guidance, satellites support climate adaptation in forest management and disaster prevention.

Energy systems are increasingly reliant on space technology to adapt to climate change. Satellites provide real-time data on solar radiation, wind patterns, and hydrological cycles, enabling renewable energy operators to optimize production. Solar and wind farms can adjust output forecasts and storage strategies based on orbital observations. Grid operators use satellite-derived data to anticipate energy demand and prevent disruptions caused by extreme weather. By integrating space technology into energy management, communities can ensure reliable power supply while minimizing reliance on fossil fuels. Space-based insights help create adaptive energy systems that are both sustainable and resilient to climate variability.

Ecosystem management is enhanced through satellite monitoring of biodiversity, habitat health, and environmental stressors. Satellites track deforestation, coral reef bleaching, and wetland degradation, providing data for conservation planning. AI algorithms analyze patterns in vegetation, water quality, and wildlife populations to predict ecological shifts. Governments and environmental organizations use these insights to implement protective measures, restore habitats, and manage natural resources sustainably. Space technology allows for continuous, large-scale monitoring of ecosystems, ensuring early detection of threats and efficient response. By applying orbital data to ecosystem management, humanity can maintain biodiversity and support climate adaptation efforts effectively.

Public health strategies also benefit from space technology in adapting to climate change. Satellites monitor temperature, humidity, and vector habitats, helping predict outbreaks of diseases influenced by environmental conditions, such as malaria, dengue, and cholera. Early warning systems informed by orbital data allow health authorities to mobilize resources, launch preventive campaigns, and target vulnerable populations. AI-driven models integrate satellite observations with local data to anticipate disease spread and optimize intervention strategies. By leveraging space-derived information, public health systems can better protect populations from climate-related health risks. Space technology bridges the gap between environmental changes and effective health planning.

Infrastructure planning and urban resilience depend on space technology for climate adaptation. Satellites provide data on flood risk, land subsidence, and storm intensity, guiding the design of roads, bridges, and buildings. AI models predict the impact of extreme weather on infrastructure, enabling engineers and planners to implement proactive measures. Cities can integrate green infrastructure, flood defenses, and energy-efficient designs informed by orbital data. This approach minimizes damage during climate events and ensures long-term urban sustainability. Space technology empowers urban planners to build resilient communities capable of withstanding climate-related challenges.

Climate policy and governance are strengthened by space-derived data. Governments and international organizations use satellite observations to assess climate vulnerabilities, prioritize adaptation projects, and monitor implementation effectiveness. Transparent, real-time data supports evidence-based policy, international cooperation, and accountability. By incorporating space technology into climate governance, decision-makers can design targeted strategies, optimize resource allocation, and track progress toward adaptation goals. Space research ensures that climate adaptation is guided by accurate information, making interventions more effective and equitable.

Early warning systems for climate-related disasters heavily rely on space technology. Satellites continuously monitor storms, hurricanes, floods, and droughts, providing data that enable predictive models and real-time alerts. AI algorithms analyze this information to forecast disaster intensity, timing, and impact, allowing governments and communities to prepare and respond effectively. Evacuation plans, emergency supplies, and infrastructure protection measures are informed by satellite observations. By leveraging space-based early warning systems, societies can reduce fatalities, property damage, and economic losses during extreme events. Space technology ensures timely, accurate, and actionable information for disaster preparedness.

Coastal management benefits from satellite-derived data in adapting to climate change. Rising sea levels, storm surges, and coastal erosion threaten communities, infrastructure, and ecosystems. Satellites provide high-resolution measurements of tides, shoreline changes, and sediment movement. Urban planners and environmental agencies use these observations to design seawalls, restore wetlands, and implement zoning regulations that protect vulnerable areas. AI modeling predicts future scenarios, guiding sustainable coastal development. By incorporating space technology into coastal adaptation, governments can balance human development with environmental preservation and resilience.

Agricultural adaptation planning is enhanced by space-based observations. Satellites monitor rainfall, soil moisture, crop health, and temperature fluctuations, enabling farmers to make data-driven decisions. AI systems analyze patterns to recommend planting schedules, irrigation strategies, and crop selection that optimize yields under changing climate conditions. Early detection of droughts, pests, and diseases allows proactive interventions, minimizing losses and ensuring food security. By integrating satellite data into agricultural management, communities can build climate-resilient farming systems. Space technology supports adaptive agriculture that responds effectively to environmental variability.

Water resource management is also strengthened through space technology. Satellites track river flows, reservoir levels, and groundwater availability, providing critical data for drought prediction and water allocation. AI-driven modeling combines these observations with local usage patterns to optimize water distribution and prevent shortages. Urban centers, agricultural areas, and industrial sectors rely on satellite data to manage water sustainably under climate stress. By leveraging space-derived insights, policymakers can prioritize interventions, reduce competition for scarce resources, and enhance resilience. Space technology ensures informed water management that adapts to climate variability.

Insurance and risk management increasingly utilize space technology for climate adaptation planning. Satellite data informs risk assessments for floods, wildfires, hurricanes, and other climate-related hazards. Insurers use AI algorithms to model potential losses, set premiums, and design mitigation strategies. Governments and private companies rely on this information to prepare for economic impacts and implement resilience measures. By integrating space-based observations into risk management, stakeholders can reduce vulnerability and improve financial and operational preparedness. Space technology strengthens climate adaptation by linking data, decision-making, and strategic planning.

The overall impact of space technology on climate adaptation is transformative, providing critical tools for resilience, sustainability, and global collaboration. Satellites offer continuous monitoring of Earth's atmosphere, oceans, land, and ecosystems, supplying real-time data for informed decision-making. AI algorithms and predictive models process this data to anticipate extreme weather, optimize resource use, and guide adaptation strategies. Communities, governments, and organizations can implement proactive measures to protect lives, infrastructure, and livelihoods from climate-related risks. By integrating space-derived insights into policy, planning, and operations, societies are better equipped to respond to the growing challenges of climate change. Space technology enables a proactive approach to adaptation rather than a reactive one.

Coastal regions, agricultural zones, and urban centers benefit from the application of orbital observations. Satellite data informs infrastructure design, disaster preparedness, crop planning, and water management. Predictive analytics allow early interventions that reduce economic losses and safeguard human health. By combining space technology with local knowledge, decision-makers can design strategies tailored to specific environmental conditions. Space-based insights ensure that adaptation measures are evidence-driven, efficient, and sustainable. The integration of space technology strengthens resilience across multiple sectors and geographies.

Public health systems are enhanced through space-derived climate monitoring. Satellites track environmental factors that influence disease outbreaks, enabling preventive interventions and resource allocation. AI-driven forecasts allow health authorities to anticipate climate-sensitive diseases, plan emergency responses, and reduce health risks. Communities benefit from improved access to preventive care and early warning systems. Space technology links environmental data with public health strategies, demonstrating the value of orbital observations for human well-being. By supporting proactive health management, space innovation contributes to climate-resilient societies.

Global collaboration is facilitated by space technology, as satellite data is often shared across borders and organizations. International partnerships enable coordinated adaptation strategies, knowledge exchange, and joint planning for climate challenges. Shared space-derived insights promote equitable resource distribution and collective action in disaster response, ecosystem management, and sustainable development. Space technology serves as a common platform that unites nations and communities in addressing climate change. By fostering cooperation, satellites and AI systems amplify the impact of adaptation measures worldwide.

Ultimately, space technology provides the tools, data, and insights necessary for humanity to adapt effectively to a changing climate. From disaster preparedness and coastal protection to agriculture, water management, and public health, satellites and space-derived systems enhance resilience, sustainability, and informed decision-making. By leveraging orbital observations, AI analytics, and predictive modeling, societies can respond proactively to climate risks, protect vulnerable populations, and build a sustainable future. Space technology demonstrates that innovations developed beyond Earth have direct, tangible benefits in supporting human adaptation and resilience on our planet.

Chapter . 27

Food Security Through Space-Based Solutions

Ensuring sufficient and equitable food distribution using satellite monitoring and space technology.

Space technology plays a vital role in supporting global food security by providing critical data for agriculture, logistics, and disaster management. Satellites monitor crop health, soil moisture, precipitation, and temperature patterns, enabling farmers to make informed decisions about planting, irrigation, and harvest timing. This real-time information allows for early detection of crop stress, disease, and pest infestations, reducing losses and maximizing yields. Governments and international organizations can use satellite data to anticipate food shortages and allocate resources efficiently. By leveraging space-based systems, societies can improve food availability, stability, and accessibility. Space technology ensures that agricultural production is optimized to meet the needs of a growing global population.

Precision agriculture benefits immensely from space-based monitoring. High-resolution satellite imagery allows farmers to identify variations in soil fertility, water availability, and crop performance across individual fields. AI-driven analysis interprets this data, recommending targeted interventions such as fertilization, irrigation, or pest control. By applying resources precisely where needed, farmers increase productivity while reducing environmental impact. Space-derived insights also support crop rotation planning, sustainable land use, and yield forecasting. These capabilities enable farmers to adapt to climate variability, ensure consistent food production, and mitigate the risk of famine. Space technology bridges the gap between data and actionable agricultural decisions.

Disaster preparedness and mitigation for agriculture rely on space technology. Satellites track droughts, floods, and extreme weather events that threaten food production. Early warning systems provide farmers, governments, and aid organizations with critical information to protect crops, livestock, and infrastructure. AI models analyze satellite data to forecast potential yield losses and guide emergency interventions. By acting proactively, communities can minimize the impact of climate-related shocks on food availability. Space-based monitoring ensures timely, evidence-based responses that safeguard both livelihoods and food security.

Global supply chains and logistics are improved through space technology. Satellite navigation and tracking systems optimize the transportation of food from farms to markets, reducing spoilage, delays, and fuel consumption. Real-time monitoring enables efficient distribution of agricultural products, even in remote or disaster-affected regions. AI systems integrate satellite data with weather forecasts, traffic conditions, and inventory levels to manage logistics dynamically. By streamlining supply chains, space technology ensures that food reaches populations in need efficiently and equitably. This contributes to both economic stability and global food security.

Policy and governance for food security are strengthened by space-based data. Governments use satellite imagery and climate models to develop strategies for sustainable agriculture, disaster response, and resource allocation. International organizations coordinate relief efforts and development programs using consistent, real-time satellite information. Transparent data sharing allows communities and stakeholders to make informed decisions about crop management, market interventions, and food distribution. By incorporating space-derived insights into policy, societies ensure that food security strategies are effective, equitable, and adaptive. Space technology provides the foundation for evidence-based decision-making in global food systems.

Case studies demonstrate how space-based solutions support food security across different regions. In India, satellites monitor monsoon patterns, soil moisture, and crop health, providing farmers with guidance to optimize planting and irrigation schedules. Early warnings of droughts or floods allow agricultural communities to protect crops and mitigate losses. Government agencies use this data to plan food distribution and emergency relief, ensuring that vulnerable populations have access to essential resources. By integrating satellite observations into agricultural planning, India improves resilience against climate variability. Space technology enables data-driven decisions that enhance productivity and safeguard livelihoods.

In Sub-Saharan Africa, satellite data is used to address chronic food insecurity caused by unpredictable weather and limited infrastructure. Satellites track rainfall, vegetation growth, and soil conditions, allowing farmers to implement adaptive practices. AI-driven analysis predicts potential crop failures and guides interventions such as supplemental irrigation, fertilizer application, or pest control. Humanitarian organizations rely on satellite-based monitoring to allocate food aid and resources efficiently during periods of drought or famine. By leveraging space-derived insights, African nations can reduce hunger, stabilize food supplies, and improve agricultural sustainability. Space technology provides critical support in regions most vulnerable to food scarcity.

Latin America benefits from space-based agricultural monitoring to optimize large-scale farming operations. Satellites provide high-resolution imagery that allows farmers to detect crop stress, monitor growth patterns, and plan harvesting schedules. AI algorithms integrate weather forecasts, soil data, and satellite imagery to recommend precise interventions that maximize yield and reduce waste. Governments and cooperatives use satellite insights to predict regional food production and ensure equitable distribution. Space technology enables efficient management of agricultural resources, helping communities adapt to climate variability while maintaining economic viability. Satellite monitoring improves both productivity and sustainability in large-scale agriculture.

Food distribution and logistics are enhanced globally through space technology. Satellites track shipments, monitor storage conditions, and optimize delivery routes to reduce spoilage and ensure timely access to markets. In disaster-affected regions, satellite navigation systems coordinate rapid food deliveries to isolated communities. AI systems analyze satellite and ground-level data to predict demand, adjust routes, and allocate resources dynamically. By improving the efficiency of food supply chains, space-based solutions ensure that both urban and rural populations receive reliable access to nutritious food. Space technology strengthens global food networks by reducing waste and enhancing distribution accuracy.

Coastal and island nations rely on satellites to support fisheries and aquaculture as part of food security. Satellites monitor ocean temperatures, currents, and algal blooms, helping manage sustainable fish stocks and predict seasonal availability. AI-driven models guide fishing schedules, optimize harvest locations, and prevent overfishing. These tools ensure that seafood resources are used efficiently and sustainably. Governments use satellite data to regulate fishing practices and support local economies dependent on marine resources. Space technology connects environmental monitoring with food production, ensuring long-term sustainability and security for marine-based communities.

Space technology plays a critical role in climate adaptation for agriculture, helping farmers anticipate and respond to environmental changes. Satellites monitor temperature fluctuations, precipitation patterns, and extreme weather events, providing farmers with timely data to adjust planting schedules and irrigation practices. AI-driven models analyze this information to predict crop yields, identify areas at risk of drought or flooding, and optimize resource use. By incorporating satellite observations into decision-making, agricultural systems become more resilient to climate variability. Governments and organizations use this data to implement policies that support sustainable farming practices and food security. Space technology enables proactive agricultural adaptation, reducing vulnerability to climate shocks.

Crop disease prediction is another key application of space-based monitoring. Satellites capture high-resolution imagery of vegetation health, detecting subtle changes in color, moisture, and growth patterns that indicate potential disease outbreaks. AI algorithms process these observations to forecast the spread of pathogens and pests, allowing early intervention before widespread crop loss occurs. Farmers can take targeted action, such as applying treatments or adjusting irrigation, minimizing economic losses and preserving yields. By leveraging space-derived data, communities reduce the impact of disease on food supply and maintain agricultural productivity. Space technology provides a powerful tool for safeguarding crops against both natural and climate-related threats.

Global monitoring for food security relies on space technology to provide comprehensive, real-time insights into agricultural productivity. Satellites track vegetation, soil moisture, rainfall, and temperature across continents, enabling international organizations to assess potential food shortages. AI-driven analytics combine satellite data with socio-economic indicators to identify regions at risk of famine or malnutrition. Policymakers use this information to coordinate international aid, plan interventions, and allocate resources effectively. By integrating space-derived insights into global monitoring, the international community can respond proactively to emerging food crises. Space technology ensures that interventions are informed, timely, and efficient.

Livestock management also benefits from satellite monitoring and AI analysis. Satellites track grazing areas, pasture health, and water availability, helping farmers manage herd health and optimize feed distribution. AI algorithms predict potential shortages or overgrazing, guiding adaptive strategies that maintain productivity while preserving the environment. Space technology allows for large-scale monitoring of livestock systems, improving efficiency, sustainability, and resilience to climate variability. Farmers can use these insights to maintain stable production and support regional food security. Space-based tools extend agricultural intelligence to both crops and livestock.

International collaboration in food security is facilitated by shared space-based data. Satellites operated by different countries and organizations provide consistent, high-resolution information accessible to governments, NGOs, and research institutions worldwide. This cooperation enables coordinated responses to food shortages, climate impacts, and disease outbreaks. By pooling resources and leveraging space-derived insights, nations can implement effective strategies for food security on a global scale. Space technology fosters collaboration, transparency, and equitable access to critical agricultural information. Shared orbital data ensures that food security solutions benefit humanity collectively.

Irrigation management is significantly enhanced through space technology, providing data for efficient water use in agriculture. Satellites monitor soil moisture, precipitation, and evaporation rates, enabling farmers to optimize irrigation schedules and conserve water. AI-driven models analyze these observations to determine where and when water is needed, preventing over-irrigation and reducing waste. This approach improves crop yields while maintaining environmental sustainability. Governments and agricultural agencies can use satellite data to develop water allocation policies and support drought-prone regions. Space technology ensures that irrigation systems are precise, efficient, and resilient to climate variability.

Soil monitoring is another critical application of space-based observations. Satellites provide information on soil composition, fertility, moisture levels, and erosion patterns across large agricultural regions. AI algorithms process this data to recommend fertilization, crop rotation, and soil conservation practices. Farmers can take targeted action to maintain soil health, reduce nutrient loss, and enhance long-term productivity. By integrating satellite insights into soil management, agricultural systems become more sustainable and resilient to climate change. Space technology offers a scalable solution for maintaining soil quality and supporting global food production.

Sustainable farming practices are facilitated by combining satellite monitoring with AI and robotics. Space-derived data guides precision agriculture techniques, such as site-specific fertilization, targeted pest control, and optimized planting patterns. Robotic systems can execute these interventions with high accuracy, reducing chemical usage and minimizing environmental impact. Farmers can maintain high yields while conserving natural resources and protecting biodiversity. By applying space technology to farming, societies can implement sustainable agricultural methods at scale. Space innovation bridges advanced technology with environmental stewardship in agriculture.

Emergency food response systems rely on space technology for rapid and effective interventions. Satellites provide real-time data on crop failures, extreme weather events, and environmental hazards affecting food production. AI-driven models assess the severity of food shortages and guide the allocation of emergency aid, including seeds, fertilizers, and food supplies. Logistic networks use satellite navigation to coordinate rapid delivery to affected regions. By leveraging space-derived insights, governments and humanitarian organizations can respond efficiently to crises, minimizing hunger and social disruption. Space technology ensures that emergency food responses are timely, precise, and effective.

Monitoring food supply chains is another application of space technology in supporting food security. Satellites track the transportation of agricultural products, storage conditions, and inventory levels. AI algorithms integrate this data with weather forecasts, demand projections, and logistics information to optimize distribution. Farmers, distributors, and policymakers can anticipate bottlenecks, reduce waste, and ensure that food reaches markets and vulnerable populations efficiently. Space technology enables a proactive approach to managing the flow of food from production to consumption. By providing real-time oversight, satellites help maintain both local and global food security.

The overall impact of space technology on global food security is transformative, providing tools for sustainable production, efficient distribution, and climate-resilient agriculture. Satellites monitor crops, soil, water, and weather patterns in real time, giving farmers, governments, and humanitarian organizations the data needed to make informed decisions. AI and predictive models process this information to forecast yields, identify risks, and optimize interventions. By integrating space-derived insights into agriculture, societies can minimize crop losses, improve resource efficiency, and respond proactively to environmental challenges. Space technology ensures that food systems are both productive and resilient to the pressures of climate change and population growth.

Global supply chains benefit from space technology by reducing waste and ensuring equitable distribution of food. Satellite tracking, navigation, and monitoring systems optimize transportation, inventory management, and logistics. AI-driven analysis predicts demand, allocates resources, and coordinates deliveries to urban centers, remote communities, and disaster-affected areas. By improving the efficiency of food distribution networks, space-based solutions ensure that populations receive timely access to nutritious food. Space technology strengthens the reliability, transparency, and sustainability of global food systems.

Climate adaptation in agriculture is enhanced through space-derived monitoring and predictive insights. Satellites provide early warnings for droughts, floods, pests, and disease outbreaks, allowing farmers to adjust practices and safeguard yields. AI algorithms translate this information into actionable guidance, supporting decision-making at farm, regional, and national levels. Space technology enables proactive adaptation, reducing vulnerability and supporting long-term food security. Communities can maintain productivity even under the pressures of climate change, ensuring consistent and stable food availability.

Policy and governance for food security are strengthened by the availability of space-based data. Governments, international organizations, and NGOs use satellite observations to prioritize interventions, allocate resources, and monitor outcomes. Transparent, real-time data supports evidence-based planning, coordinated responses, and equitable food distribution. By integrating space technology into policy frameworks, decision-makers can address hunger, manage risks, and promote sustainable agricultural practices. Space technology provides a foundation for informed, accountable, and effective governance of food systems worldwide.

Ultimately, space technology demonstrates that solutions developed for monitoring and managing Earth from orbit can have direct, tangible benefits for food security. From precision agriculture and soil monitoring to disaster response, supply chain optimization, and policy planning, satellites and space-derived tools ensure that humanity can meet the growing demand for food sustainably and equitably. By leveraging orbital insights, predictive analytics, and AI-driven guidance, societies can build resilient, adaptive, and sustainable food systems. Space technology proves that innovation beyond Earth can safeguard life on Earth itself, securing food for current and future generations.

Chapter . 28

Global Transportation and Logistics Optimization

Improving supply chains and reducing carbon footprints with space navigation systems and satellite monitoring.

Space technology plays a transformative role in optimizing global transportation and logistics systems. Satellites provide real-time monitoring of shipping routes, flight paths, and cargo movements, allowing for precise tracking and coordination. GPS navigation, initially developed for space and military applications, enables accurate positioning and route planning for trucks, ships, and aircraft. AI-driven algorithms analyze satellite data to optimize delivery schedules, reduce fuel consumption, and minimize delays. By integrating space technology into logistics, companies can streamline supply chains, improve efficiency, and reduce operational costs. Space-based insights ensure that goods move more predictably and sustainably across global networks.

Maritime shipping benefits significantly from satellite monitoring. Ocean-going vessels rely on GPS, satellite imagery, and automated tracking systems to navigate complex shipping lanes safely. Satellites provide data on weather conditions, sea currents, and potential hazards such as icebergs or piracy zones. AI and predictive models use this information to optimize routes, reduce travel time, and minimize fuel consumption. By applying space technology, shipping companies can enhance safety, improve reliability, and reduce carbon emissions. Space-derived tools allow maritime transport to be more efficient, environmentally responsible, and resilient to unforeseen disruptions.

Air transportation is similarly improved through space-based systems. Aircraft rely on GPS, satellite-based communication, and weather monitoring to optimize flight paths and manage air traffic. Satellite data enables airlines to avoid turbulent zones, plan fuel-efficient routes, and respond to changing weather conditions in real time. AI-driven analytics integrate flight schedules, passenger demand, and airport operations to improve efficiency and reduce delays. By leveraging space technology, the aviation industry can enhance safety, reduce environmental impact, and increase reliability in global air travel. Satellites provide the framework for precise, real-time navigation and operational optimization.

Land-based logistics also benefit from satellite-enabled monitoring and AI optimization. Delivery trucks, freight trains, and distribution networks rely on GPS and real-time traffic data to plan efficient routes and reduce congestion. AI systems analyze satellite imagery and sensor data to anticipate delays, optimize scheduling, and improve supply chain coordination. Logistics companies can reduce fuel use, carbon emissions, and operational costs while ensuring timely delivery of goods. By applying space-based navigation and monitoring, land transportation becomes more reliable, sustainable, and responsive. Space technology transforms terrestrial logistics by integrating precise data and intelligent decision-making.

Disaster response and humanitarian logistics are strengthened through space technology. Satellites provide real-time imagery of disaster-affected areas, helping coordinate the delivery of emergency supplies, food, and medical aid. AI algorithms analyze terrain, road conditions, and infrastructure damage to identify the fastest and safest delivery routes. By using space-derived data, governments and aid organizations can respond quickly and efficiently to emergencies, reducing human suffering and resource loss. Space technology ensures that logistics operations remain effective even under extreme conditions. This capability demonstrates the vital role of orbital insights in supporting global transportation and humanitarian resilience.

Case studies highlight the transformative impact of space technology on global logistics. In Europe, railway networks use GPS and satellite monitoring to track trains, optimize scheduling, and reduce energy consumption. AI systems analyze real-time data on train locations, track conditions, and passenger demand to improve efficiency and minimize delays. Similarly, freight and delivery companies rely on satellite-based tracking to monitor vehicle performance, predict maintenance needs, and reduce downtime. By integrating space-derived insights, European logistics networks achieve higher reliability, lower operational costs, and improved sustainability. Space technology enables precise, data-driven decision-making across complex transportation systems.

In North America, trucking and delivery companies leverage satellite navigation and AI optimization for long-haul freight transport. GPS allows trucks to follow the most efficient routes, avoid traffic congestion, and reduce fuel consumption. AI algorithms integrate weather forecasts, traffic conditions, and delivery schedules to adjust routes dynamically. Fleet managers can monitor vehicle performance, anticipate maintenance needs, and ensure timely delivery of goods across vast distances. By combining space technology with intelligent logistics management, companies reduce emissions, save costs, and enhance reliability. Satellite-enabled solutions improve operational efficiency while supporting environmental sustainability.

Air cargo operations around the world also rely heavily on space technology. Airlines use satellite-based communication, GPS, and weather monitoring to plan optimal flight paths, minimize fuel consumption, and maintain schedules. AI models analyze real-time data on air traffic, cargo volumes, and demand fluctuations to optimize routing and maximize capacity. During emergencies, satellites provide critical situational awareness for rerouting flights and delivering urgent goods efficiently. By leveraging space-derived insights, the aviation industry enhances safety, reliability, and environmental performance. Space technology transforms air cargo into a precise, responsive, and sustainable system.

Maritime logistics benefit from satellite-enabled monitoring and predictive analytics. Shipping companies use satellite imagery to track port congestion, ocean currents, and weather events. AI systems analyze this information to optimize vessel speed, routes, and fuel use. Real-time monitoring reduces the risk of delays, accidents, and cargo damage. Coastal and remote regions gain improved access to essential goods through optimized shipping operations. By integrating space technology, maritime supply chains become more efficient, environmentally responsible, and resilient to disruptions. Satellites provide the infrastructure for global trade to operate reliably under dynamic conditions.

Sustainability in logistics is enhanced through space-based insights. Companies can calculate and reduce carbon footprints by analyzing satellite data on fuel consumption, route efficiency, and environmental impact. AI-driven optimization ensures that vehicles, ships, and aircraft use energy effectively, minimizing greenhouse gas emissions. Governments and organizations can monitor compliance with environmental regulations and assess sustainability metrics at a global scale. Space technology provides transparency and actionable intelligence, enabling logistics operations to balance efficiency, reliability, and ecological responsibility. Satellites ensure that transportation networks evolve in harmony with climate and sustainability goals.

Emergency response logistics rely heavily on space technology to deliver supplies efficiently during crises. Satellites provide real-time imagery of disaster-affected areas, including damaged roads, bridges, and infrastructure. AI-driven analysis of this data helps planners identify safe and efficient routes for transporting food, medical supplies, and essential equipment. GPS navigation ensures that vehicles reach affected communities on time, while satellite communication systems coordinate efforts between multiple agencies. By integrating space-derived insights, emergency logistics operations become faster, safer, and more effective. Space technology allows humanitarian organizations to respond proactively to disasters, minimizing loss of life and economic damage.

Urban delivery systems are enhanced through satellite navigation, real-time traffic monitoring, and AI optimization. In densely populated cities, delivery fleets use GPS and satellite data to avoid congestion, reduce delays, and plan efficient routes. AI algorithms integrate information on road conditions, weather, and demand fluctuations to dynamically adjust schedules. Autonomous delivery vehicles and drones, guided by space-derived positioning systems, further improve speed and accuracy. By leveraging space technology, urban logistics networks become more reliable, reduce fuel consumption, and lower emissions. Efficient urban delivery systems contribute to economic growth and environmental sustainability simultaneously.

Predictive maintenance of transportation infrastructure and vehicles benefits from space technology. Satellites monitor bridges, highways, railways, and ports, detecting structural stress, deformation, or damage caused by environmental factors. AI systems analyze satellite imagery and sensor data to predict when maintenance is needed, preventing accidents and reducing downtime. Fleet operators use space-based monitoring to track vehicle health, anticipate component failures, and schedule timely repairs. By applying these insights, transportation networks maintain high reliability, minimize operational disruptions, and extend the lifespan of critical infrastructure. Space technology enables a proactive approach to maintenance and safety.

Multi-modal transportation networks are optimized using space-based data and AI. Integrating shipping, rail, road, and air transport requires precise coordination and real-time monitoring. Satellites provide the geospatial data necessary to track cargo, optimize transfer points, and manage inventory across modes of transport. AI algorithms model demand, route efficiency, and environmental impact to determine the best logistical strategies. By leveraging space-derived insights, organizations can create seamless, interconnected supply chains that reduce costs, improve reliability, and minimize carbon emissions. Space technology facilitates intelligent management of complex, global transportation systems.

Global trade and commerce are strengthened through the integration of space technology into logistics. Satellites provide transparency in tracking goods from production to consumption, allowing companies to verify delivery, assess risks, and respond to disruptions. AI-driven models analyze trade flows, port activity, and environmental conditions to optimize operations and prevent bottlenecks. By using space-based tools, businesses can enhance competitiveness, maintain supply chain continuity, and support sustainable practices. Space technology ensures that global commerce operates efficiently, reliably, and with reduced environmental impact. The integration of orbital data into logistics exemplifies how space innovation directly improves Earth-based economic systems.

Case studies demonstrate the effectiveness of space technology in optimizing global logistics and transportation systems. In Japan, rail networks use GPS and satellite-based monitoring to coordinate high-speed train schedules, minimize delays, and optimize energy consumption. AI systems analyze real-time data on track conditions, passenger flow, and train locations to adjust operations dynamically. This approach reduces operational costs, improves efficiency, and enhances safety. By integrating space technology into railway logistics, Japan achieves reliable and sustainable transportation for millions of passengers daily. Satellite-based solutions enable predictive planning and seamless coordination across complex networks.

In the United States, trucking and delivery companies leverage satellite navigation and monitoring to improve supply chain efficiency. GPS tracking, real-time traffic data, and AI route optimization help fleets avoid congestion, reduce fuel consumption, and meet tight delivery schedules. Logistics managers can monitor vehicle performance, anticipate maintenance needs, and adjust operations proactively. Space-based tools also support disaster response, enabling rapid distribution of emergency supplies in hurricane-affected regions. By combining orbital data with intelligent management systems, the U.S.

transportation sector enhances reliability, reduces emissions, and ensures timely delivery of goods nationwide.

Maritime shipping demonstrates the global benefits of space technology for logistics. Shipping companies rely on satellites to track vessels, monitor ocean currents, and avoid extreme weather. AI models use this data to optimize travel routes, reduce fuel usage, and improve delivery timelines. Ports use satellite imagery to manage congestion, coordinate docking schedules, and streamline cargo handling. By integrating space-derived insights, maritime logistics networks operate more efficiently, safely, and sustainably. Satellite monitoring enables proactive decision-making that reduces environmental impact while maintaining global trade flow.

Air cargo operations also benefit from space technology in optimizing logistics. Airlines utilize satellite communication, GPS navigation, and weather monitoring to plan fuel-efficient flight paths and maintain precise delivery schedules. AI-driven systems integrate real-time cargo demand, aircraft availability, and route conditions to maximize capacity and reduce delays. Emergency air transport relies on satellites to coordinate rapid responses to crises, delivering critical supplies and medical aid efficiently. By applying space technology, aviation logistics becomes more reliable, environmentally responsible, and responsive to dynamic conditions. Satellites provide the foundation for a modern, efficient air cargo system.

Humanitarian logistics and emergency supply chains are strengthened through space technology. Satellites provide real-time imagery and geospatial data for areas affected by natural disasters, conflict, or food shortages. AI analysis identifies optimal delivery routes, predicts infrastructure challenges, and coordinates multi-agency responses. GPS and satellite communications ensure that supplies reach remote or disrupted regions quickly and safely. By leveraging space-based tools, humanitarian organizations can improve response speed, efficiency, and resource allocation. Space technology demonstrates its critical role in ensuring equitable access to essential goods during emergencies worldwide.

The overall impact of space technology on global transportation and logistics is profound, providing tools for efficiency, sustainability, and resilience. Satellites and space-based navigation systems enable precise monitoring of cargo, vehicles, and infrastructure, allowing supply chains to operate smoothly across continents. AI-driven analysis of satellite data optimizes routes, predicts disruptions, and ensures timely delivery of goods. By integrating space technology into logistics, organizations can reduce operational costs, minimize carbon emissions, and improve overall system reliability. Space-derived insights ensure that goods move efficiently, safely, and predictably from production to consumption.

Maritime, aviation, rail, and trucking industries all benefit from orbital data and space-enabled tools. Shipping companies use satellite tracking to optimize routes and reduce fuel consumption. Airlines rely on satellite-based weather monitoring and navigation to plan fuel-efficient flight paths. Rail networks leverage GPS and AI-driven analytics to improve scheduling and infrastructure management. Trucking and delivery fleets use satellites to avoid congestion, predict maintenance needs, and coordinate routes. Across all modes of transport, space technology enhances efficiency, reduces environmental impact, and increases reliability.

Emergency and humanitarian logistics are strengthened through space technology, ensuring that supplies reach those in need during disasters and crises. Satellites provide real-time imagery of affected regions, allowing rapid assessment of infrastructure damage and optimal route planning. AI-driven systems coordinate the distribution of food, medicine, and essential resources, reducing delays and improving outcomes. Space-based communication ensures coordination between multiple agencies and remote regions. By applying these tools, humanitarian organizations can respond more effectively and save lives. Space technology proves essential in building resilient logistics systems capable of operating under extreme conditions.

Sustainability is another key advantage of space-enabled logistics. Satellite monitoring and AI optimization reduce fuel consumption, greenhouse gas emissions, and environmental degradation associated with transportation networks. Real-time tracking and predictive analytics allow organizations to minimize waste, improve energy efficiency, and support global climate goals. By incorporating space-derived insights, industries can balance operational efficiency with ecological responsibility. Space technology ensures that logistics networks evolve in a way that is both economically and environmentally sustainable.

Ultimately, space technology transforms global transportation and logistics into intelligent, adaptive systems. Satellites, AI, and GPS navigation provide the data, insights, and coordination necessary for efficient, reliable, and sustainable supply chains. From everyday commercial deliveries to emergency humanitarian operations, space-based tools optimize operations, reduce environmental impact, and enhance resilience. By leveraging orbital observations, predictive models, and advanced analytics, humanity can maintain robust global transportation networks that support economic growth, social equity, and environmental stewardship. Space innovation demonstrates its direct and essential impact on the movement of goods and services across Earth.

Chapter . 29

Space Technology in Renewable Resource Management

Utilizing satellites to optimize forests, water, and energy usage sustainably.

Space technology is revolutionizing the management of renewable resources by providing precise, real-time data on forests, water systems, and renewable energy sources. Satellites monitor the health and coverage of forests, track changes in water bodies, and measure solar and wind energy potential across regions. AI-driven models analyze this data to guide sustainable management, optimize resource use, and reduce environmental impact. Governments, industries, and conservation organizations rely on space-derived insights to make informed decisions, allocate resources efficiently, and implement long-term sustainability plans. By leveraging space technology, societies can ensure that renewable resources are used responsibly while meeting growing human needs. Satellites provide a comprehensive view of Earth's natural assets, enabling smarter stewardship and planning.

Forest management benefits significantly from space-based monitoring. Satellites provide detailed imagery of forest cover, deforestation rates, and biodiversity hotspots. AI algorithms detect early signs of illegal logging, forest degradation, and fire risks, allowing for rapid intervention. Forest managers can optimize reforestation efforts, assess carbon sequestration, and maintain ecosystem services. By integrating satellite data, policies for forest conservation and sustainable timber production are more accurate and effective. Space technology ensures that forests are managed to balance economic, ecological, and social needs. This contributes directly to climate mitigation and the preservation of biodiversity.

Water resource management is another area enhanced by space technology. Satellites track river flows, reservoir levels, groundwater availability, and soil moisture across regions. AI models integrate these observations to forecast water scarcity, plan irrigation, and optimize allocation for agriculture, industry, and domestic use. By monitoring water distribution in real time, governments can prevent shortages, manage drought conditions, and protect aquatic ecosystems. Space technology allows for proactive management, reducing conflicts over water resources and ensuring long-term sustainability. By combining orbital data with predictive analytics, communities can maintain secure and equitable access to freshwater.

Renewable energy development is optimized through space-derived data. Satellites measure solar irradiance, wind patterns, and other environmental variables essential for planning solar farms, wind turbines, and hydroelectric systems. AI models analyze this information to determine the most efficient placement and operation of renewable energy infrastructure. By leveraging space technology, energy producers can maximize output, reduce costs, and minimize environmental disruption. Governments can use this data to develop energy policies, support grid management, and transition toward low-carbon power systems. Space-based insights ensure that renewable energy systems are designed and operated for efficiency, sustainability, and resilience.

Sustainable agriculture is further supported through satellite monitoring of land, water, and climate conditions. By providing data on soil fertility, precipitation patterns, and vegetation health, satellites enable precise management of crops and pasturelands. AI-driven models recommend adaptive strategies that optimize water usage, reduce chemical inputs, and increase productivity. This approach ensures that food production aligns with environmental conservation and resource sustainability goals. By integrating space technology into agriculture, communities can maintain high yields while preserving natural resources. Satellites provide the foundation for managing renewable resources in a way that balances human needs with ecological integrity.

Case studies illustrate how space technology supports sustainable management of renewable resources worldwide. In Brazil, satellites monitor the Amazon rainforest, tracking deforestation, forest degradation, and reforestation efforts. AI algorithms detect illegal logging and predict areas at risk, enabling rapid interventions by government agencies and environmental organizations. By using space-derived data, Brazil can protect biodiversity, maintain carbon sequestration, and support sustainable timber production. Satellite monitoring ensures that forests are managed efficiently and responsibly, balancing ecological conservation with economic needs. Space technology provides the tools to maintain forest health on a massive scale.

In Australia, satellite imagery supports water resource management in arid and semi-arid regions. Satellites track river flows, groundwater levels, and soil moisture, allowing authorities to allocate water efficiently for agriculture, urban supply, and ecosystem preservation. AI-driven models predict drought conditions, optimize irrigation schedules, and identify potential water shortages before they become critical. By leveraging space technology, Australia can manage its limited water resources sustainably and reduce conflicts over water allocation. Satellite monitoring ensures that water systems remain resilient under changing climate conditions, supporting both human and ecological needs.

Renewable energy planning benefits from space technology in Europe. Satellites provide high-resolution data on solar irradiance, wind patterns, and cloud coverage across regions, enabling precise siting of solar panels and wind turbines. AI models integrate this information with electricity demand, grid capacity, and environmental constraints to optimize energy production and distribution. By using space-derived data, European nations can maximize renewable energy efficiency, reduce reliance on fossil fuels, and lower carbon emissions. Satellite monitoring ensures that renewable energy projects are both economically viable and environmentally sustainable. Space technology supports the transition to clean energy on a continental scale.

Agricultural sustainability is enhanced through satellite monitoring in Africa. Satellites track soil health, rainfall patterns, and vegetation growth, providing farmers with insights to optimize planting schedules, irrigation, and fertilization. AI-driven analysis identifies areas at risk of drought, pest infestations, or nutrient depletion, allowing for targeted interventions. By integrating space-based insights, African farmers can increase yields while minimizing environmental impact. Satellite monitoring supports sustainable land use, protects natural resources, and strengthens food security across vulnerable regions. Space technology provides the knowledge necessary for climate-smart agriculture.

Coastal and marine resource management is strengthened through satellites. In Southeast Asia, satellites monitor mangroves, coral reefs, and fisheries, providing data for sustainable management of marine ecosystems. AI-driven analysis predicts overfishing, habitat degradation, and environmental risks, guiding policy and conservation efforts. By using space technology, governments can protect biodiversity, maintain fish stocks, and ensure that coastal communities benefit from sustainable resource use. Satellite monitoring provides actionable insights that balance human activity with environmental preservation. Space-based systems make it possible to manage renewable resources effectively across land, water, and marine ecosystems.

Energy systems benefit greatly from space technology, allowing for optimized production, distribution, and consumption of renewable resources. Satellites provide continuous data on solar irradiance, wind speeds, and hydrological cycles, helping energy producers predict output and plan grid management. AI-driven models integrate this data with energy demand forecasts to optimize generation, reduce waste, and prevent shortages. By using space-based insights, renewable energy systems become more efficient, reliable, and resilient to climate variability. Governments and utilities can make informed decisions regarding energy infrastructure, ensuring sustainability and reducing environmental impact. Space technology bridges observation, prediction, and action in energy resource management.

Resource scarcity and potential shortages can be mitigated through satellite monitoring and predictive analytics. Satellites track water levels, soil moisture, forest health, and crop conditions, providing early warnings of potential shortages or stress. AI algorithms analyze trends to predict future resource availability and guide strategic planning. Governments and organizations can implement adaptive measures to prevent crises in water, food, or energy supplies. Space technology ensures that decisions are based on accurate, real-time, and large-scale data. By anticipating resource scarcity, societies can maintain stability and sustainability.

Policy planning for renewable resource management is strengthened by space-based data. Governments, NGOs, and international bodies use satellite observations to assess the health of forests, water systems, and energy infrastructure. Transparent, real-time data informs regulations, conservation strategies, and sustainable development initiatives. Satellite monitoring allows for accurate tracking of resource use, compliance with environmental policies, and assessment of the effectiveness of interventions. By integrating space technology into policy, decision-makers can ensure that renewable resources are managed responsibly and equitably. Space-derived insights support governance that balances human needs with environmental stewardship.

Global collaboration is facilitated through shared space technology and data. Countries and organizations can exchange satellite observations and AI models to coordinate conservation efforts, optimize renewable energy systems, and manage transboundary resources. International collaboration enables early detection of environmental risks, equitable resource allocation, and synchronized emergency response to natural disasters. By using space-based tools collectively, nations can address sustainability challenges that extend beyond borders. Space technology acts as a platform for cooperation, transparency, and shared responsibility in renewable resource management.

Predictive modeling enhances long-term sustainability planning by combining satellite data, AI analysis, and climate projections. Governments, industries, and conservation organizations can forecast potential resource stress, optimize land and water use, and evaluate the impact of renewable energy projects. Predictive insights help allocate investments efficiently, minimize environmental damage, and enhance resilience to climate change. By leveraging space technology, societies can adopt proactive strategies rather than reactive measures. Satellites and AI enable informed, forward-looking management of Earth's renewable resources. Space-derived predictive models ensure that sustainability initiatives remain effective under evolving environmental conditions.

Case studies demonstrate the tangible benefits of space technology in managing renewable resources. In Canada, satellites monitor vast forested areas, detecting signs of pest infestations, fire risks, and illegal logging. AI systems analyze these observations to guide forest management decisions, reforestation efforts, and conservation policies. By integrating space-derived data, Canadian authorities maintain forest health, protect biodiversity, and optimize carbon sequestration. Satellite monitoring ensures that forests are managed sustainably while balancing economic and ecological priorities. Space technology provides a comprehensive view that enables large-scale environmental stewardship.

Water conservation projects in California leverage satellite imagery to manage reservoirs, rivers, and irrigation networks. Satellites provide real-time data on soil moisture, water levels, and precipitation, allowing water managers to allocate resources efficiently. AI-driven models predict droughts, optimize irrigation schedules, and identify areas of water stress. By using space technology, California can maintain reliable water supply for agriculture, industry, and urban areas. Satellite monitoring enables proactive management, reducing waste and preventing shortages. Space-derived insights support sustainable water use in regions facing climate variability.

Renewable energy planning in Germany benefits from satellite and AI integration. Satellites measure solar irradiance, wind patterns, and cloud coverage to determine optimal locations for solar and wind farms. Predictive models estimate energy output, guide infrastructure investment, and improve grid integration. By leveraging space technology, Germany maximizes renewable energy efficiency, reduces dependency on fossil fuels, and minimizes environmental impact. Satellite monitoring ensures that energy systems are designed and operated sustainably. Space technology provides critical support for achieving national and global energy transition goals.

Integrated resource management is exemplified by the European Union's approach to agriculture, forestry, and water systems. Satellites provide consistent, high-resolution data across multiple sectors, allowing policymakers to coordinate efforts for maximum efficiency and sustainability. AI models analyze trends, assess environmental impacts, and optimize resource allocation. By using space technology, the EU can maintain ecological balance, support renewable resource use, and ensure food and energy security. Satellite-derived insights enable cross-sector collaboration and long-term strategic planning. Space technology integrates multiple data streams to manage natural resources holistically.

In Africa, space technology supports sustainable land use and ecosystem management. Satellites monitor soil conditions, vegetation growth, and water availability across agricultural regions. AI-driven analytics help farmers optimize planting schedules, irrigation, and crop rotation to increase yield and conserve resources. Governments and NGOs use satellite data to develop policies for sustainable agriculture, water management, and forest protection. By applying space-derived insights, African nations enhance food security while preserving natural resources for future generations. Space technology empowers communities to manage renewable resources efficiently and responsibly.

The overall impact of space technology on renewable resource management is profound, enabling more efficient, sustainable, and resilient use of forests, water, and energy systems. Satellites provide real-time monitoring of natural resources, while AI-driven models interpret these data to optimize allocation, prevent waste, and mitigate environmental risks. Governments, industries, and communities can make informed decisions that balance human needs with ecological sustainability. By integrating space-derived insights, renewable resource management becomes proactive rather than reactive, ensuring long-term availability and stability. Space technology allows for large-scale oversight and coordination, making sustainability achievable at regional, national, and global levels.

Forest management benefits directly from satellite monitoring and predictive analysis. High-resolution imagery detects deforestation, illegal logging, and forest degradation, allowing rapid intervention and sustainable planning. AI models optimize reforestation projects, assess carbon sequestration, and maintain biodiversity. Space technology ensures that forest ecosystems continue to provide critical ecological services while supporting economic needs. By enabling precise, data-driven management, satellites strengthen global climate mitigation efforts and biodiversity preservation. Forests become a managed resource that aligns human activity with environmental responsibility.

Water resource management is also transformed through space-based insights. Satellites track rivers, lakes, groundwater, and soil moisture to monitor availability and predict shortages. AI-driven predictive models guide irrigation, urban water distribution, and emergency planning. By leveraging satellite data, water managers can respond proactively to droughts, optimize usage, and prevent conflicts over scarce resources. Space technology ensures that water systems remain reliable, resilient, and sustainable under climate variability. Communities benefit from equitable access to water while conserving vital freshwater ecosystems.

Renewable energy development is optimized using satellite monitoring and analysis. Satellites measure solar, wind, and hydrological patterns, guiding the placement and operation of energy infrastructure. Predictive AI models ensure maximum efficiency, reduced waste, and minimal environmental impact.

Governments and energy producers can integrate space-derived insights into policy and grid management, advancing clean energy adoption. By leveraging space technology, renewable energy systems become more reliable, sustainable, and cost-effective. Satellites enable a global transition toward low-carbon energy while supporting economic growth.

Ultimately, space technology provides the tools, insights, and predictive capabilities necessary for sustainable, integrated management of renewable resources. From forests and water systems to energy production and agricultural land, satellites and AI ensure that natural resources are used responsibly and equitably. By combining orbital monitoring, predictive modeling, and international collaboration, societies can achieve long-term sustainability and resilience. Space technology demonstrates that innovation beyond Earth can have direct, tangible benefits in managing our planet's resources. Renewable resource management becomes smarter, more efficient, and globally coordinated through space-based solutions.

Chapter . 30

Empowering Developing Nations Through Space Access

Providing satellite tools to enable growth and resilience in emerging economies.

Access to space technology offers developing nations unprecedented opportunities to accelerate economic growth, improve governance, and enhance resilience. Satellites provide vital data on agriculture, natural resources, climate patterns, and infrastructure, enabling evidence-based decision-making. Communication satellites bridge the digital divide, delivering internet access to remote and underserved areas. By leveraging space-based tools, developing countries can improve education, healthcare, and disaster preparedness. Space technology empowers nations to participate in the global economy while managing their unique environmental and social challenges. Providing access to orbital data and satellite systems ensures that developing nations are not left behind in technological advancement.

Agricultural productivity is significantly enhanced through space technology. Satellites monitor crop health, soil conditions, water availability, and weather patterns, enabling farmers to optimize planting, irrigation, and harvesting. AI-driven analytics translate satellite observations into actionable guidance, helping smallholder farmers improve yields and reduce losses. Governments can use these insights to manage food security, plan resource allocation, and prevent famine. By integrating space-based tools into agriculture, developing nations increase food production sustainably while protecting natural resources. Space technology allows communities to adapt to climate variability and mitigate the impact of environmental stressors. Improved agricultural efficiency contributes to economic stability and social resilience.

Disaster preparedness and response are strengthened through space access. Satellites provide real-time data on earthquakes, floods, hurricanes, and other hazards, allowing early warnings and rapid intervention. AI and satellite analytics identify vulnerable regions, optimize evacuation routes, and guide the distribution of emergency aid. Developing nations can use these capabilities to minimize loss of life, reduce infrastructure damage, and allocate resources efficiently. Space technology transforms reactive disaster management into proactive planning, increasing community resilience. By ensuring access to critical satellite data, emerging economies can better protect citizens from natural and human-made disasters. Space-based solutions make humanitarian response faster, smarter, and more coordinated.

Education and skill development benefit from exposure to space technology. Satellites enable remote learning, provide internet access in rural areas, and facilitate STEM education initiatives. Developing nations can use space programs to inspire innovation, train scientists and engineers, and build local technical expertise. Access to space technology encourages collaboration with international research institutions and fosters a culture of technological advancement. By integrating satellites and space-derived data into educational programs, nations empower future generations to lead in science, engineering, and policy-making. Space access becomes a catalyst for human capital development and innovation-driven growth.

Healthcare systems are enhanced through satellite-enabled data and communication networks. Satellites provide telemedicine capabilities, track disease outbreaks, and monitor environmental factors that affect public health. AI models use space-based observations to predict and manage epidemics, allocate medical resources, and guide vaccination campaigns. Developing nations can implement more effective and timely healthcare interventions using space technology. Satellite connectivity ensures that remote and underserved communities receive medical support, reducing health disparities. Space technology strengthens public health infrastructure while improving resilience to environmental and social challenges.

Real-world examples illustrate the transformative impact of space technology on developing nations. In Kenya, satellites are used to monitor rainfall patterns, soil moisture, and crop health, enabling farmers to optimize planting schedules and irrigation. AI-driven models interpret this data, helping smallholder farmers increase productivity and reduce losses from drought or pests. Government agencies use satellite insights to manage national food security and allocate resources more efficiently. By leveraging space technology, Kenya can enhance agricultural resilience, ensure food availability, and support rural livelihoods. Satellites provide the foundation for sustainable, climate-smart farming practices.

In Bangladesh, satellites support disaster management and flood prediction. Real-time satellite imagery tracks river levels, storm surges, and rainfall patterns, allowing authorities to issue early warnings and coordinate evacuations. AI models analyze satellite data to identify the most vulnerable communities and predict areas at risk of flooding or landslides. By integrating space-based insights into emergency planning, Bangladesh can minimize loss of life and reduce infrastructure damage. Satellite communications ensure timely coordination between government agencies, NGOs, and first responders. Space technology empowers developing nations to transform disaster response from reactive to proactive.

Education in remote regions of India is enhanced through satellite-enabled internet and communication systems. Satellites provide connectivity to rural schools, enabling access to online learning platforms, STEM programs, and virtual classrooms. Students and teachers gain exposure to advanced scientific and technical resources, expanding educational opportunities. Satellite-based programs also facilitate training in space science, robotics, and data analytics, building local expertise. By integrating space technology into education, India can reduce the urban-rural learning gap and empower students to participate in the knowledge economy. Access to space-based tools fosters innovation, critical thinking, and long-term human capital development.

Healthcare delivery is improved in countries such as Nigeria through satellite-supported telemedicine and disease monitoring. Satellites enable remote consultations, real-time communication between hospitals, and the tracking of infectious diseases. AI algorithms analyze satellite and health data to predict outbreaks, optimize resource allocation, and guide vaccination campaigns. By leveraging space technology, Nigeria can improve healthcare accessibility for remote communities and respond efficiently to public health challenges. Satellites enhance monitoring of environmental factors that influence health, such as air quality, water availability, and temperature fluctuations. Space technology strengthens healthcare systems, saving lives and improving overall population health.

Infrastructure planning and urban development in emerging economies are supported by satellite data. Cities in Southeast Asia use satellite imagery to monitor land use, population growth, and transportation networks. AI-driven models analyze this data to guide urban planning, disaster preparedness, and sustainable infrastructure projects. Developing nations can optimize resource allocation, reduce environmental impact, and improve resilience to climate change. By integrating space technology into policy and planning, governments can design smarter, safer, and more equitable cities. Satellites provide the data needed to balance rapid urbanization with environmental sustainability and social well-being.

Governance in developing nations is strengthened through space technology, enabling data-driven decision-making and improved policy implementation. Satellites provide accurate, real-time information on natural resources, population distribution, and infrastructure, allowing governments to plan effectively and monitor compliance. AI and geospatial analytics turn satellite observations into actionable insights for resource allocation, environmental protection, and urban planning. By leveraging space technology, policymakers can reduce inefficiencies, prevent corruption, and ensure equitable access to public services. Transparent satellite data enhances accountability and supports evidence-based governance. Space technology empowers developing nations to create efficient, informed, and responsive government systems.

Economic development is accelerated through access to space-based tools. Satellites provide critical data for agriculture, fisheries, energy, and transportation, enabling businesses to optimize operations and reduce costs. AI-driven analysis of satellite data identifies market trends, natural resource availability, and logistical opportunities. By integrating space technology into commercial sectors, emerging economies can enhance productivity, attract investment, and create new employment opportunities. Small and medium enterprises benefit from improved supply chain management, crop monitoring, and infrastructure planning. Space-based tools allow developing nations to compete in global markets and stimulate sustainable economic growth. Access to orbital technology becomes a lever for innovation and prosperity.

Climate monitoring and adaptation are made more effective with space technology. Satellites track temperature changes, rainfall, soil moisture, and deforestation, providing essential data to predict environmental risks. AI-driven models use this information to guide disaster preparedness, agricultural planning, and water resource management. Developing nations can implement climate adaptation strategies to reduce vulnerability to floods, droughts, and extreme weather events. By leveraging satellite insights, countries can protect ecosystems, maintain food security, and safeguard communities. Space technology enables evidence-based climate action that builds resilience and reduces economic losses. Monitoring Earth from orbit helps nations anticipate environmental challenges before they escalate.

International collaboration is enhanced through shared access to space-based tools and data. Developing nations can partner with global organizations, space agencies, and universities to gain technical expertise, funding, and satellite imagery. Collaborative initiatives allow emerging economies to participate in global monitoring programs, share best practices, and receive training in space science and AI analytics. By working together, nations can address cross-border challenges such as climate change, deforestation, and natural disasters. Space technology fosters equitable participation in technological innovation and global problem-solving. Access to satellites and data enables developing nations to strengthen resilience and achieve sustainable development goals collectively.

Innovation and entrepreneurship flourish when developing nations have access to space technology. Local businesses and startups can leverage satellite data to create solutions in agriculture, renewable energy, transportation, and environmental monitoring. AI and geospatial analytics enable tailored services that meet local needs while addressing global challenges. By providing space-derived tools, governments and international organizations empower citizens to develop technology-driven solutions that improve livelihoods. Access to space technology becomes a catalyst for creativity, competitiveness, and technological self-reliance. Emerging economies can harness satellites and orbital data to transform societal challenges into opportunities for growth.

Case studies illustrate how developing nations effectively leverage space technology to improve livelihoods and resilience. In Rwanda, satellites monitor agricultural productivity, soil health, and rainfall, enabling farmers to plan crops and irrigation efficiently. AI-driven analytics provide actionable insights, allowing smallholder farmers to maximize yield while conserving resources. Government agencies use this data to allocate food aid, plan agricultural subsidies, and strengthen food security. Space technology empowers Rwanda to reduce vulnerability to climate variability while enhancing economic growth. Satellites provide the critical intelligence necessary for informed, sustainable agricultural management.

In the Philippines, satellite-based disaster monitoring enhances emergency response and disaster preparedness. Satellites track typhoons, floods, and landslides in real time, providing early warnings to communities and authorities. AI models analyze this data to predict the severity and impact of disasters, enabling rapid resource deployment and evacuation planning. By integrating space technology into disaster management systems, the Philippines reduces fatalities, minimizes property damage, and coordinates aid efficiently. Satellite communications ensure that remote and vulnerable areas receive timely alerts and assistance. Space technology transforms disaster response from reactive to proactive, saving lives and resources.

Education initiatives in Nepal benefit from satellite-enabled internet access and digital learning platforms. Remote communities use satellites to access online classrooms, STEM programs, and virtual libraries. Students and teachers gain exposure to advanced scientific knowledge, space science, and technological tools. By integrating space technology into education, Nepal reduces educational disparities between urban and rural regions and fosters innovation. Satellites allow students to develop skills that prepare them for careers in science, technology, and engineering. Space access provides the foundation for knowledge-based development and human capital growth.

Healthcare delivery in Bangladesh is strengthened through satellite-supported telemedicine and disease surveillance. Satellites provide connectivity for remote clinics, track epidemic outbreaks, and monitor environmental factors affecting health. AI-driven analysis of satellite data supports early detection of disease clusters, resource allocation, and targeted vaccination campaigns. By leveraging space technology, healthcare systems become more responsive, efficient, and equitable. Satellite monitoring ensures that vulnerable populations in rural and disaster-prone regions receive timely medical support. Space-based tools improve public health outcomes while enhancing community resilience.

Infrastructure planning in Vietnam benefits from satellite imagery and geospatial analytics. Satellites monitor urban growth, road networks, water systems, and coastal areas, enabling efficient planning and disaster risk mitigation. AI-driven models analyze this data to optimize land use, transportation networks, and sustainable development projects. By integrating space technology,

Vietnamese authorities can balance rapid urbanization with environmental conservation and resilience planning. Satellite monitoring ensures that infrastructure projects are efficient, safe, and sustainable. Space-based tools provide emerging economies with the insights needed to build smarter, more resilient cities.

The overall impact of space technology on empowering developing nations is transformative, providing tools to drive economic growth, social development, and resilience. Satellites and space-derived data enable governments to make informed decisions in agriculture, disaster management, healthcare, and infrastructure planning. AI-driven analysis of this data supports predictive modeling, risk mitigation, and efficient resource allocation. By integrating space technology, emerging economies can reduce vulnerability to climate change, natural disasters, and food insecurity. Access to orbital tools ensures that nations participate equitably in global technological advancement. Space technology becomes a catalyst for sustainable development and social progress.

Agricultural systems benefit directly from satellite monitoring and predictive insights. Farmers gain access to accurate data on soil health, crop growth, and water availability, allowing for precision agriculture and improved yields. Governments can implement policies to secure food supplies, allocate resources effectively, and prevent famine. By leveraging space-based tools, developing nations strengthen food security while maintaining environmental sustainability. Satellites empower communities to adapt to changing climate conditions and improve resilience against agricultural challenges. Space technology ensures that food production supports both human needs and ecological stewardship.

Disaster management is significantly enhanced through space technology. Satellites provide real-time monitoring of floods, storms, earthquakes, and other hazards, enabling early warnings and rapid response. AI analysis predicts vulnerable areas and guides resource deployment, evacuation plans, and relief efforts. By applying space technology, developing nations reduce human and economic losses during crises and improve preparedness for future events. Satellite communication ensures coordination between agencies, NGOs, and affected populations. Space-derived data transforms disaster response from reactive to proactive, increasing the safety and resilience of communities.

Education, healthcare, and infrastructure development are strengthened through access to space technology. Satellites provide connectivity for remote learning, telemedicine, and data-driven urban planning. AI and geospatial analytics enable governments and organizations to optimize resource allocation, improve public services, and foster innovation. By leveraging space-based tools, emerging economies can close gaps in education and healthcare access, build resilient infrastructure, and stimulate technological advancement. Space technology empowers nations to nurture human capital and strengthen social systems. Satellites provide the platform for sustainable and inclusive development.

Ultimately, access to space technology enables developing nations to participate fully in global innovation while addressing local challenges. From agriculture and disaster preparedness to education, healthcare, and infrastructure, satellites and AI provide actionable insights that drive growth, resilience, and equity. By integrating space-derived tools into governance, policy, and industry, emerging economies can achieve sustainable development goals, reduce vulnerability, and improve quality of life for citizens. Space technology demonstrates that participation in the orbital economy is not only a path to technological advancement but a means to empower nations, strengthen societies, and ensure a more equitable world.

Part VII: The Future of Space Technology for Earth

Chapter . 31

Space Mining and Earth Resource Management

Exploring how extracting resources from space could reduce pressure on terrestrial ecosystems.

Space mining has the potential to revolutionize resource management on Earth by providing access to minerals, metals, and rare elements found in asteroids and other celestial bodies. Satellites and robotic spacecraft can survey, identify, and extract these materials without causing ecological damage on Earth. AI-driven systems analyze orbital and surface data to determine resource-rich locations and optimize extraction processes. By sourcing critical resources from space, humanity can reduce the environmental impact of mining, preserve natural habitats, and prevent overexploitation of terrestrial ecosystems. Space mining allows for sustainable resource acquisition while supporting industrial, technological, and energy needs. The integration of orbital technology with Earth-based industries ensures responsible, forward-looking resource management.

Rare earth elements, essential for electronics, renewable energy systems, and advanced technology, can be sourced from space, alleviating the strain on terrestrial reserves. Satellites and remote sensing systems identify asteroids rich in elements like platinum, nickel, and cobalt. Robotic mining operations in orbit or on asteroid surfaces extract materials with minimal environmental footprint. By supplementing Earth's supply, space mining reduces the need for destructive mining practices, lowering habitat destruction, soil erosion, and water pollution. This approach ensures a sustainable supply of critical materials for emerging technologies. Space-based resource management supports both industrial growth and environmental stewardship.

Water and volatiles found on the Moon or asteroids can also be harvested to support Earth and space operations. Ice deposits on celestial bodies can provide drinking water, agricultural irrigation, or be converted into hydrogen and oxygen for fuel. Satellites monitor these resources, mapping their location and quantity for efficient extraction. By sourcing water and volatiles from space, developing nations and industries can reduce pressure on freshwater ecosystems and support sustainable development. Space mining allows humanity to expand resource access while conserving critical Earth-based water supplies. Satellites and AI provide the precision needed to manage these extraterrestrial resources responsibly.

Energy-related materials, such as helium-3 and rare isotopes, can also be mined from celestial bodies for use in clean energy technologies. Helium-3, for example, is a potential fuel for nuclear fusion, offering nearly limitless energy with minimal waste. Satellites monitor lunar surfaces or asteroid belts to locate high concentrations of these materials. By supplementing energy resources from space, Earth-based mining and fossil fuel extraction can be reduced, lowering environmental degradation and greenhouse gas emissions. Space mining provides a pathway to sustainable energy solutions and long-term resource security. Satellites and orbital systems enable informed decision-making for safe and efficient extraction.

Space mining also supports circular economy initiatives on Earth by providing raw materials for manufacturing without exhausting terrestrial resources. Metals and minerals from asteroids can feed high-tech industries, construction, and renewable energy systems. By sourcing materials from space, supply chains become more sustainable and resilient to geopolitical or environmental disruptions. AI-driven planning ensures efficient extraction, transport, and utilization of space-mined resources. Integrating space mining with Earth-based resource management reduces ecological strain while meeting global industrial demands. Space technology demonstrates a pathway to balance economic growth with environmental preservation.

Case studies illustrate how space mining could transform Earth's resource management and reduce environmental strain. Companies and space agencies are conducting surveys of near-Earth asteroids to identify mineral-rich targets, including platinum-group metals, nickel, and cobalt. AI and satellite data help pinpoint asteroids with the highest economic potential, optimizing mission planning and resource allocation. By extracting these materials in space, Earth-based mining operations can be reduced, protecting ecosystems, forests, and water sources from degradation. Countries and industries can secure a steady supply of critical resources without causing ecological harm. Space mining represents a paradigm shift in balancing industrial needs with environmental responsibility.

Technological requirements for space mining are advancing rapidly. Robotic spacecraft, autonomous mining systems, and AI-driven resource mapping enable efficient extraction from asteroids, the Moon, or other celestial bodies. Satellites provide high-resolution imagery, spectroscopic data, and orbital analysis to guide robotic operations. Energy systems for extraction, including solar and nuclear-powered units, allow mining missions to operate sustainably in space. By leveraging cutting-edge technology, space mining becomes feasible, safe, and economically viable. These tools also minimize risk to humans and reduce the environmental footprint of traditional mining practices. Space technology provides a blueprint for sustainable resource acquisition beyond Earth.

Economic models indicate that space mining could stabilize global supply chains and reduce dependence on politically or environmentally sensitive terrestrial mines. Rare metals and minerals obtained from asteroids can be delivered to Earth to support electronics, renewable energy, and high-tech industries. By supplementing Earth's resource base, space mining reduces scarcity, price volatility, and geopolitical tensions over critical materials. AI-driven market simulations can forecast demand and optimize resource distribution, ensuring that extraction benefits multiple sectors and nations. This approach allows both developed and developing nations to access essential materials without compromising Earth's ecosystems. Space technology facilitates economic growth while promoting environmental sustainability.

Environmental benefits of space mining are substantial. By shifting mining operations off Earth, natural habitats are preserved, water and soil pollution is minimized, and greenhouse gas emissions from extraction and transportation are reduced. Forests, rivers, and wetlands are spared from destructive practices, allowing ecosystems to thrive. Satellite monitoring ensures that Earth-based resource use is sustainable, while orbital operations supplement industrial demand. Space mining demonstrates that humanity can meet growing technological needs without depleting or damaging terrestrial ecosystems. This approach provides a model for responsible resource management that aligns economic and environmental goals.

Legal and ethical frameworks are critical to ensure that space mining benefits humanity equitably. International agreements, regulatory bodies, and transparent resource-sharing policies must guide extraction and utilization. Satellites and AI monitoring can ensure compliance with treaties and environmental guidelines. By establishing ethical practices, developing and developed nations alike can participate in space mining responsibly. Space-based resource management becomes a tool for global cooperation, economic stability, and environmental stewardship. Proper governance ensures that orbital mining supports both human development and planetary preservation.

Transporting space-mined resources to Earth requires innovative logistical solutions to ensure efficiency, safety, and economic viability. Satellites and orbital monitoring systems track the trajectory of mined materials, guiding their safe return to the planet's surface. Advanced propulsion systems, autonomous navigation, and AI-driven orbital calculations minimize the risk of collisions and optimize fuel consumption. Re-entry capsules are designed to protect the materials while ensuring controlled delivery to designated processing facilities. By coordinating space-based mining and transportation, industries can integrate extraterrestrial resources seamlessly into terrestrial supply chains. Space technology ensures that materials reach Earth safely while reducing the environmental impact of transport.

Integration with terrestrial industries is key to maximizing the benefits of space-mined resources. Metals and minerals from asteroids can support electronics manufacturing, renewable energy technologies, and advanced construction materials. AI-driven predictive models match supply with demand, optimizing distribution and reducing waste. By incorporating space-derived resources into existing industrial processes, companies can reduce dependence on environmentally destructive mining operations. Satellite data and AI analytics help track materials, manage logistics, and ensure responsible use. This integration strengthens economic resilience while supporting sustainability objectives. Space mining allows Earth industries to thrive without exhausting natural reserves.

Resource recycling and circular economy practices are enhanced by space mining. By supplementing terrestrial resources with materials from space, industries can reduce overexploitation of Earth's finite reserves. Satellites monitor both space and Earth operations, providing insights for efficient resource use and minimizing waste. AI algorithms predict when materials will be needed, facilitating proactive recycling and reuse. Integrating space-mined resources into circular economy frameworks supports sustainable industrial growth. Space technology ensures that resource management is holistic, efficient, and environmentally responsible. By combining orbital and terrestrial strategies, humanity can meet industrial demands while protecting ecosystems.

Long-term sustainability strategies rely on combining space mining with renewable energy and environmental monitoring. Solar-powered mining units, AI-driven operations, and satellite communication networks reduce the carbon footprint of extraction.

Space-mined resources enable the development of clean energy technologies, reducing reliance on fossil fuels and lowering greenhouse gas emissions. Environmental monitoring from satellites ensures that Earth's ecosystems remain protected as industrial activity increases. By planning sustainably, space mining supports global development goals while preserving planetary health. Space technology enables a balanced approach where resource extraction benefits society without compromising the environment.

Global cooperation is essential to ensure the equitable and responsible use of space resources. International agreements, treaties, and regulatory frameworks govern ownership, extraction, and distribution. Satellites provide transparency, monitoring adherence to agreements and promoting accountability. Collaborative programs allow nations to share technology, expertise, and benefits, ensuring that space mining contributes to global prosperity. By fostering partnerships and ethical practices, space technology strengthens international relations and resource equity. Responsible space mining demonstrates that humanity can explore, extract, and innovate while protecting both Earth and outer space.

Space mining presents environmental, economic, and technological challenges that must be addressed for long-term success. One concern is the risk of space debris generated by mining operations, which could threaten satellites, spacecraft, and orbital infrastructure. AI-driven monitoring and collision-avoidance systems, along with precise orbital planning, help mitigate these risks. Economic challenges include the high cost of launching, operating, and maintaining mining missions, which require significant investment and international collaboration. Technological challenges involve robotics, autonomous systems, and materials processing in extreme space environments. By addressing these issues, space mining can become a sustainable and safe solution for Earth's resource demands. Space technology provides the tools to overcome these challenges through careful planning and innovation.

Environmental considerations on Earth remain a priority, even with space-mined resources. By reducing terrestrial mining, space mining lowers habitat destruction, water contamination, and greenhouse gas emissions. Satellites monitor mining operations and material transport, ensuring minimal impact on ecosystems during integration with terrestrial industries. AI analytics help forecast resource demand and optimize supply chains, preventing overexploitation and waste. By combining orbital extraction with Earth-based environmental management, humanity can balance industrial growth with ecological preservation. Space technology allows for resource acquisition without compromising planetary health.

Economic models for space mining are evolving to support global collaboration and investment. Public-private partnerships, international consortia, and joint ventures allow nations to share costs, expertise, and benefits. Satellites provide real-time data for investors, governments, and companies to make informed decisions regarding resource allocation and mission planning. AI-driven financial modeling predicts market demand, pricing, and supply chain needs, ensuring profitable and sustainable operations. By integrating economic planning with space technology, nations can leverage space mining to strengthen industrial capacity and reduce dependence on environmentally sensitive terrestrial resources. Space mining creates opportunities for equitable and sustainable economic growth.

Technological innovations are critical for the success of space mining initiatives. Robotic miners, AI-driven navigation, autonomous resource processing, and advanced propulsion systems all contribute to efficient extraction and transport. Satellites provide precise mapping, monitoring, and communication between Earth and mining operations. Materials processing in space reduces contamination and energy consumption compared to Earth-based mining. By advancing technology, space mining becomes feasible, safe, and scalable. Space-based innovations also drive progress in robotics, AI, and remote sensing for other applications on Earth.

Space technology ensures that challenges are met with adaptive and forward-thinking solutions.

Case examples illustrate the practical applications and potential of space mining. NASA, ESA, and private companies are actively developing asteroid mining missions to extract metals and volatiles for use on Earth and in space exploration. These projects integrate satellite monitoring, AI analysis, and autonomous robotics to identify, extract, and transport resources efficiently. Lessons learned from these initiatives inform future strategies, reduce risks, and optimize mission design. By studying and implementing these examples, nations can plan sustainable and responsible space mining operations. Satellites provide the intelligence and oversight needed to make space mining an environmentally friendly and economically viable solution.

The overall impact of space mining on Earth's resource management is profound, offering a sustainable solution to the growing demand for critical minerals, metals, and energy resources. By extracting materials from asteroids, the Moon, and other celestial bodies, humanity can reduce the ecological and social costs of terrestrial mining. Satellites and AI-driven systems ensure precise monitoring, safe transport, and efficient integration of space-mined resources into Earth's industries. By supplementing terrestrial supplies, space mining alleviates pressure on forests, rivers, and soils, preserving ecosystems for future generations. Space technology provides the foundation for responsible resource acquisition that balances industrial growth with environmental stewardship.

Rare earth elements, metals, and volatiles obtained from space are essential for electronics, renewable energy, and advanced manufacturing. By sourcing these materials from orbit, industries can maintain innovation and production without harming Earth's ecosystems. AI and satellite monitoring ensure that extraction and transport operations are safe, efficient, and environmentally responsible. Space mining demonstrates that economic growth and ecological preservation can coexist when technology is applied thoughtfully. Satellites provide real-time intelligence, enabling proactive management of both extraterrestrial and terrestrial resources. Space technology transforms resource scarcity into an opportunity for sustainable development.

Energy security benefits from space-mined resources as well. Helium-3, water ice, and other space-derived materials can support nuclear fusion, fuel production, and renewable energy systems on Earth. By providing alternative energy sources, space mining reduces reliance on fossil fuels, lowering greenhouse gas emissions and mitigating climate change. Satellite monitoring ensures optimal resource utilization and integration with terrestrial energy infrastructure. Space technology enables a transition toward cleaner, sustainable energy while supporting industrial and societal needs. The combination of space resources and Earth-based innovation fosters long-term energy resilience.

Global collaboration is essential for realizing the full potential of space mining. International agreements, regulatory frameworks, and shared technology initiatives ensure equitable access, transparency, and responsible management of extraterrestrial resources. Satellites monitor compliance, track resource movement, and provide data for cooperative planning. By fostering collaboration, space mining becomes a model for sustainable development that benefits all nations. Space technology enables nations to work together, balancing economic, environmental, and ethical considerations in a rapidly evolving sector.

Ultimately, space mining provides a pathway to sustainable resource management that safeguards Earth's ecosystems while supporting economic growth and technological advancement. Satellites, AI, and robotic mining systems allow humanity to access essential materials without overexploiting terrestrial resources. By integrating space-derived resources into global industries, we reduce environmental damage, ensure long-term availability, and create new opportunities for innovation. Space technology demonstrates that exploration beyond Earth can directly benefit life on our planet. Responsible space mining ensures a future where human development and environmental preservation advance together.

Chapter . 32

Global Security and Safety via Space Surveillance

Using satellites to enhance security without escalating terrestrial conflict.

Space surveillance has become a critical tool for global security, offering real-time monitoring of military, environmental, and strategic activities from orbit. Satellites provide continuous observation of borders, maritime routes, and airspace, ensuring that nations have accurate information to prevent conflict. AI-driven analytics interpret satellite imagery, detect anomalies, and predict potential security threats, allowing proactive decision-making. By using space surveillance, countries can maintain security while reducing the need for aggressive on-the-ground or naval operations. Space-based monitoring enables transparency and accountability, fostering trust between nations. The integration of orbital technology into security strategies ensures that protection measures are precise, efficient, and non-intrusive.

Maritime security is strengthened through satellite monitoring of shipping lanes, ports, and international waters. Satellites track vessel movements, detect unauthorized activities, and provide early warnings of piracy, smuggling, or illegal fishing. AI models analyze patterns of movement to identify unusual or potentially dangerous behavior. By using space-based surveillance, nations can protect trade routes, maintain economic stability, and ensure compliance with international laws. Space technology allows monitoring without deploying large fleets or increasing military presence, reducing the risk of escalation. Satellites provide accurate, timely information that supports maritime safety and global commerce.

Airspace monitoring is enhanced by space surveillance systems, which track aircraft movements and identify potential threats. Satellites detect unregistered flights, monitor border violations, and provide intelligence to civil and military aviation authorities. AI algorithms analyze flight paths and detect unusual patterns, allowing rapid response to potential security incidents. By integrating orbital observation into airspace management, nations can ensure safety while minimizing conflict risks. Space surveillance enables countries to monitor activity without violating sovereignty, fostering trust and transparency. Satellites provide a comprehensive and impartial view of global air traffic for security purposes.

Land and border security are supported through satellite imaging and geospatial analytics. Satellites detect illegal crossings, smuggling routes, and potential military threats along borders. AI-driven systems analyze patterns to predict and prevent incursions, allowing law enforcement and military forces to respond efficiently. By leveraging space technology, nations can reduce physical patrol costs, minimize confrontation risks, and maintain peaceful border management. Space surveillance offers a neutral, observational perspective that enhances security without direct interference. Satellite intelligence ensures timely, informed, and coordinated action to safeguard territorial integrity.

Disaster monitoring and response also contribute to global safety. Satellites detect environmental hazards, chemical spills, and nuclear incidents, providing early warnings and data for mitigation efforts. AI-driven analysis predicts the spread of hazards, enabling evacuation plans and emergency resource allocation. By combining security and environmental monitoring, space surveillance ensures public safety while supporting global stability. Countries can act collaboratively, using shared satellite data to prevent crises and minimize human and economic losses. Space technology demonstrates that security encompasses not only defense against conflict but also protection of people and ecosystems worldwide.

Case studies illustrate the practical impact of space surveillance on global security and safety. In Europe, the European Space Agency (ESA) uses satellites to monitor border regions, track migration flows, and detect illegal activities. AI-driven models analyze satellite data to identify irregular patterns, enabling authorities to respond efficiently and humanely. This approach improves security while minimizing the risk of conflict with neighboring countries. By combining real-time monitoring with predictive analytics, Europe can maintain stable borders and ensure public safety. Space surveillance provides the intelligence necessary for informed decision-making and proactive security measures.

In the United States, the National Reconnaissance Office (NRO) and other agencies leverage space-based systems to monitor military activity and safeguard critical infrastructure. Satellites track missile launches, troop movements, and cybersecurity threats, providing early warnings to national defense authorities. AI algorithms process vast amounts of data to detect anomalies and predict potential threats before they escalate. By using space surveillance, the U.S. can strengthen national security while reducing the need for costly and potentially provocative military deployments. Satellite monitoring ensures timely, accurate, and actionable intelligence for strategic decision-making. Space technology enhances both deterrence and global stability.

Maritime security operations in the Asia-Pacific region are enhanced through space surveillance. Countries like Japan, South Korea, and Australia utilize satellites to monitor shipping lanes, detect illegal fishing, and track potential threats in international waters. AI systems analyze satellite imagery to identify suspicious vessel activity and predict possible maritime hazards. By using space-based monitoring, nations can ensure safe navigation, protect trade, and maintain regional stability without escalating military presence. Satellites provide a neutral observational platform that supports cooperation among neighboring nations.

Space technology enables proactive management of maritime security challenges.

Disaster prevention and response in Southeast Asia rely on satellite monitoring to enhance safety and resilience. Satellites track typhoons, floods, volcanic activity, and landslides, providing early warnings to governments and local authorities. AI-driven analytics predict the severity and impact of disasters, enabling timely evacuations, resource allocation, and emergency response. By integrating space surveillance with disaster management systems, countries reduce fatalities, property damage, and economic disruption. Satellite communication networks also ensure coordination across agencies and regions. Space technology allows nations to combine security and safety efforts, protecting both people and infrastructure from natural and human-made hazards.

International cooperation in space surveillance fosters global security and transparency. The United Nations Office for Outer Space Affairs (UNOOSA) and other organizations facilitate data sharing, joint monitoring initiatives, and collaborative risk assessment. Satellites provide impartial observation of sensitive areas, helping nations build trust and avoid conflicts. AI-driven analysis ensures that shared data is accurate, timely, and actionable. By leveraging space technology collectively, countries can enhance security, manage natural disasters, and promote peace without relying solely on military interventions. Space surveillance demonstrates that technological collaboration can strengthen global stability and safety.

Technological systems underpinning space surveillance are critical for monitoring global security and safety. Satellites equipped with high-resolution optical and radar sensors provide continuous observation of Earth's surface, airspace, and maritime zones. AI algorithms process massive volumes of data, detecting anomalies, tracking movements, and predicting potential threats. Communication satellites relay information in real time, ensuring that authorities receive accurate intelligence quickly. Advanced geospatial analytics allow nations to map trends, identify patterns, and prioritize interventions. By combining satellite technology with AI, space surveillance transforms raw data into actionable insights for effective decision-making.

Predictive analytics is a cornerstone of modern space-based security systems. Machine learning models analyze historical and real-time satellite data to forecast potential security threats, natural disasters, or infrastructure vulnerabilities. For example, AI can detect unusual troop movements near borders, identify changes in shipping patterns, or anticipate environmental hazards. By integrating predictive models, nations can implement preventive measures, reducing response times and mitigating risks before crises escalate. Satellites provide the essential observational data, while AI interprets and contextualizes it for strategic planning. This synergy enhances national and global security without requiring aggressive or intrusive actions. Space technology enables foresight in both human and environmental safety.

Real-time monitoring ensures immediate awareness of critical events across the globe. Satellites track natural disasters, maritime incidents, airspace violations, and infrastructure threats, transmitting data instantly to control centers. AI-driven systems analyze incoming information and trigger alerts for rapid response. Governments, humanitarian organizations, and international agencies can coordinate efforts efficiently, minimizing damage and protecting populations. Real-time space surveillance allows decision-makers to act with confidence, relying on verified and current data. By providing continuous observation, satellites reduce uncertainty, improve situational awareness, and enhance security. Space technology transforms global monitoring from reactive observation to proactive management.

Integration of satellite data with terrestrial systems strengthens operational efficiency and coordination. Ground-based sensors, radar networks, and communication infrastructure complement space surveillance, providing a comprehensive security network. AI platforms synthesize inputs from multiple sources, creating accurate situational maps for military, civil, and humanitarian applications. By linking space-based and ground-based systems, countries can detect threats, deploy resources, and mitigate hazards more effectively. Satellite monitoring ensures that information is accurate, timely, and actionable. Space technology bridges observation, analysis, and response for enhanced global safety.

Ethical and legal considerations are essential for responsible space surveillance. Satellites provide an observational advantage, but transparency and international agreements are necessary to prevent misuse or escalation of tensions. Monitoring should prioritize safety, disaster prevention, and environmental protection, rather than intrusive surveillance or aggressive posturing. AI systems must be designed with accountability, fairness, and human oversight to avoid unintended consequences. By establishing clear norms and treaties, space surveillance can enhance security while respecting sovereignty and global ethics. Space technology demonstrates that global safety can be achieved through observation, intelligence, and cooperation rather than conflict.

Space surveillance has proven critical in preventing natural disasters and enhancing emergency response. In Japan, satellites monitor typhoon formation, intensity, and trajectory, providing early warnings to authorities and communities. AI-driven models analyze satellite data to predict flooding, landslides, and infrastructure risks. By integrating space-based monitoring with local emergency services, Japan can evacuate vulnerable populations, allocate resources efficiently, and reduce fatalities. Satellites ensure real-time communication and coordination during crises, enabling precise, timely interventions. Space technology transforms disaster management into a proactive system that saves lives and minimizes damage.

Border security is strengthened through satellite observation in Europe and North America. Satellites monitor remote and sensitive regions, detecting unauthorized crossings, smuggling routes, and suspicious activity. AI analytics interpret satellite imagery to identify patterns and predict potential breaches. By using space surveillance, authorities can allocate resources efficiently without escalating tensions or deploying excessive military force. Satellite data provides transparency and accountability, reducing misunderstandings between neighboring nations. Space technology ensures borders are monitored effectively while maintaining peaceful and cooperative relations.

Maritime monitoring has been enhanced significantly by space-based systems. In the South China Sea, satellites track commercial shipping, illegal fishing, and potential piracy. AI algorithms analyze vessel movements, detect anomalies, and generate early warnings for maritime authorities. By using orbital observation, countries can safeguard trade routes, protect marine ecosystems, and respond to threats without large naval deployments. Satellite monitoring ensures accurate, impartial information, fostering regional cooperation and reducing conflict risks. Space technology provides a neutral platform for overseeing international waters while supporting economic and environmental security.

International cooperative programs demonstrate the benefits of shared space surveillance. The United Nations and regional alliances coordinate satellite data sharing for disaster management, climate monitoring, and security applications. Participating nations access high-resolution imagery, predictive analytics, and communication tools to enhance preparedness and response. Shared space intelligence builds trust, reduces the potential for conflict, and facilitates coordinated action during emergencies. By pooling resources and knowledge, countries can achieve higher security and safety standards than they could individually. Satellites serve as instruments of cooperation, promoting peace and stability worldwide.

Space surveillance also contributes to monitoring critical infrastructure and environmental hazards. Satellites track the condition of dams, bridges, power grids, and industrial sites, detecting potential failures or risks. AI systems analyze changes over time to predict problems and guide preventive maintenance. By integrating space-derived data with national security and environmental policies, governments can prevent accidents, mitigate damage, and protect human lives. Space technology provides an impartial and comprehensive perspective that complements ground-based monitoring systems. Surveillance from orbit ensures that both human and ecological safety are prioritized in national and global planning.

The overall impact of space surveillance on global security and safety is transformative, providing nations with accurate, real-time intelligence to prevent conflicts and protect populations. Satellites monitor borders, airspace, maritime zones, and critical infrastructure, enabling proactive measures rather than reactive responses. AI-driven analysis converts massive volumes of satellite data into actionable insights, guiding decision-making for governments, military agencies, and humanitarian organizations. By reducing reliance on intrusive or aggressive monitoring, space technology fosters transparency, trust, and international cooperation. Space surveillance demonstrates that security can be maintained without escalating terrestrial conflict.

Disaster prevention is enhanced through orbital observation and predictive analytics. Satellites track natural hazards, monitor climate events, and detect environmental risks, providing early warnings to vulnerable regions. AI algorithms forecast the intensity and impact of storms, floods, and earthquakes, allowing authorities to plan evacuations and allocate resources efficiently. By combining global observation with predictive modeling, space surveillance saves lives, reduces property damage, and strengthens resilience. Satellite communication networks ensure coordination across regions and agencies, improving response times and effectiveness. Space technology transforms disaster management from reactive crisis response into strategic preparedness.

Maritime and border security are strengthened while minimizing confrontation and escalation. Satellites track vessel and aircraft movements, detect anomalies, and ensure compliance with international laws. AI systems analyze patterns to predict illegal activities, smuggling, and potential threats. By using space-based surveillance, nations can protect trade routes, monitor borders, and ensure territorial integrity without aggressive military action. Transparency provided by satellite monitoring fosters trust between neighboring nations and reduces the likelihood of misunderstandings. Space technology ensures effective, impartial, and non-intrusive security operations.

International cooperation is enhanced through shared space surveillance initiatives. Organizations such as the United Nations facilitate the exchange of satellite data for disaster management, climate monitoring, and security purposes. Collaborative programs allow countries to pool expertise, share resources, and establish global standards for responsible monitoring. Satellites provide a neutral platform for observation, building trust and enabling joint action to address security and environmental challenges. Space technology demonstrates that collaboration, rather than competition, strengthens global safety and resilience.

Ultimately, space surveillance ensures that global security and safety are maintained in a transparent, efficient, and ethical manner. Satellites, AI, and communication networks provide accurate, timely, and actionable information for governments, humanitarian agencies, and international organizations. By integrating orbital monitoring with predictive analytics and collaborative frameworks, nations can prevent conflicts, respond effectively to disasters, and protect critical infrastructure. Space technology demonstrates that observation, intelligence, and cooperation can replace aggression and uncertainty in maintaining security. The result is a safer, more stable, and resilient world for all humanity.

Chapter . 33

Mitigating Climate Risks with Advanced Orbital Technologies

Employing cutting-edge space solutions to anticipate and mitigate climate emergencies.

Advanced orbital technologies provide an unprecedented perspective on Earth's climate, enabling precise monitoring, forecasting, and risk mitigation. Satellites track temperature fluctuations, greenhouse gas emissions, ice coverage, sea level rise, and extreme weather patterns with high-resolution sensors. AI and machine learning analyze vast volumes of satellite data to detect trends, predict hazards, and model potential climate scenarios. By providing actionable insights, these technologies allow governments, industries, and communities to prepare for and respond to climate risks effectively. Space-based solutions help reduce the human and economic costs of climate change while supporting long-term sustainability. Integrating orbital technology into climate strategy ensures that mitigation and adaptation efforts are data-driven and proactive.

Monitoring greenhouse gases is a critical component of climate risk mitigation. Satellites equipped with spectrometers can measure carbon dioxide, methane, and other pollutants with high accuracy. AI-driven analytics identify emission hotspots, track changes over time, and predict future concentrations. By understanding sources and patterns of greenhouse gas emissions, policymakers can implement targeted mitigation strategies. Space-based monitoring ensures transparency, accountability, and international collaboration in reducing climate impact. These observations help nations meet climate agreements while safeguarding ecosystems and human health. Advanced orbital technologies transform climate policy from reactive to informed and strategic.

Predicting extreme weather events is another key application of orbital technologies. Satellites monitor atmospheric conditions, ocean temperatures, and storm formation in real time, providing early warnings for hurricanes, cyclones, floods, and droughts. Machine learning models analyze this data to forecast intensity, trajectory, and potential impact, allowing timely interventions. By using space-based prediction systems, communities can evacuate at-risk areas, secure infrastructure, and deploy emergency resources efficiently. Real-time monitoring reduces human casualties and economic losses caused by natural disasters. Space technology enhances resilience and preparedness by offering accurate, actionable forecasts. Satellites bridge the gap between climate observation and practical response strategies.

Sea level rise and polar ice monitoring are critical for understanding long-term climate risks. Satellites equipped with radar altimeters and imaging systems measure glacial retreat, ice sheet thickness, and ocean surface changes. AI models analyze trends to predict future sea level changes, helping coastal nations plan adaptation strategies. Space-based observations inform urban planning, flood defense construction, and coastal ecosystem protection. By providing precise data, orbital technologies support proactive risk management and infrastructure resilience. Advanced monitoring ensures that climate adaptation measures are timely, effective, and scientifically grounded. Space technology enables nations to anticipate climate-related threats before they become emergencies.

Agriculture and water resource management benefit significantly from orbital climate monitoring. Satellites track rainfall, soil moisture, temperature variations, and drought conditions across regions. AI-driven systems analyze these data points to optimize irrigation, planting schedules, and crop selection. By integrating satellite insights, farmers and governments can maintain food security while reducing water wastage and environmental impact. Space-based technologies enable climate-smart agriculture and sustainable water management, crucial for regions facing growing climate pressures. Advanced orbital tools ensure that agriculture adapts dynamically to changing climate conditions. Climate risk mitigation becomes both actionable and measurable through space-based monitoring.

Case studies demonstrate the practical impact of orbital technologies in mitigating climate risks. In the United States, NASA's Earth-observing satellites monitor hurricanes, wildfires, and drought conditions. AI algorithms analyze satellite data to predict storm paths, identify areas at risk of wildfire, and forecast water shortages. By providing early warnings, these systems allow authorities to evacuate communities, allocate resources, and minimize economic losses. Space-based monitoring ensures that climate adaptation strategies are informed, timely, and effective. Satellites enable decision-makers to respond proactively rather than reactively to climate hazards. Advanced orbital technology strengthens national resilience against climate-related disasters.

In Europe, the Copernicus program uses satellites to monitor greenhouse gas emissions, deforestation, and ocean health. High-resolution imagery and AI-driven analytics track environmental changes and identify areas most vulnerable to climate impacts. Governments use this information to enforce environmental regulations, implement carbon reduction strategies, and protect ecosystems. By integrating orbital data into policy, Europe can reduce climate risks while promoting sustainable development. Satellites provide a neutral, consistent, and scientifically validated source of information for climate governance. Space technology ensures that mitigation efforts are coordinated, measurable, and evidence-based. Orbital monitoring supports climate resilience at regional and continental scales.

Developing nations also benefit from climate-focused orbital technologies. In India, satellites track monsoon patterns, river flows, and agricultural conditions to inform water management and disaster preparedness. AI models process these data to predict floods, droughts, and crop failures, allowing early interventions. By combining space-based observation with local knowledge, governments and communities can protect livelihoods, maintain food security, and reduce disaster losses. Satellites provide access to real-time, accurate information that would otherwise be difficult or impossible to obtain. Space technology empowers developing nations to respond proactively to climate risks. Orbital monitoring ensures that climate adaptation is inclusive, data-driven, and scalable.

Sea-level rise monitoring is crucial for coastal regions worldwide. Satellites measure ocean height, coastal erosion, and ice sheet mass loss, providing precise data for risk assessment. AI algorithms model the impact of rising seas on communities, infrastructure, and ecosystems. Coastal planners use these insights to design seawalls, manage wetlands, and relocate vulnerable populations proactively. By integrating space technology into coastal management, nations can mitigate the most severe consequences of climate change. Satellite monitoring ensures that adaptation strategies are informed by accurate, high-resolution data. Orbital technologies provide a critical tool for preserving human and ecological well-being in vulnerable coastal zones.

International collaboration enhances the effectiveness of space-based climate risk mitigation. Programs like the UN's International Charter on Space and Major Disasters allow nations to share satellite data during emergencies. Countries can pool resources, expertise, and monitoring capabilities to respond to floods, hurricanes, and wildfires more efficiently. AI-driven systems integrate multisource data to improve predictive models and guide interventions. By fostering cooperation, orbital technologies amplify their impact and ensure that even countries with limited resources can protect citizens and ecosystems. Space technology becomes a unifying tool for global climate resilience. Collaboration ensures that climate risks are addressed collectively, scientifically, and proactively.

Satellite-based monitoring of energy systems is essential for climate risk mitigation. Satellites track the performance of solar farms, wind turbines, and hydropower installations, detecting inefficiencies and environmental impacts. AI-driven models analyze weather patterns and energy output to optimize renewable energy production and grid management. By integrating space-based observation, energy providers can maximize efficiency, reduce reliance on fossil fuels, and anticipate disruptions caused by extreme weather. Satellite data also informs infrastructure planning for resilient energy distribution in climate-vulnerable regions. Space technology ensures that energy systems adapt dynamically to changing environmental conditions, reducing greenhouse gas emissions and strengthening sustainability.

Agriculture benefits significantly from orbital climate monitoring. Satellites track soil moisture, temperature variations, crop growth, and drought conditions across vast regions. AI algorithms analyze these data points to optimize irrigation, planting schedules, and harvest timing, reducing crop loss and conserving resources. By providing real-time information, satellites allow farmers to implement climate-smart practices that improve food security and reduce environmental impact. Governments and agricultural agencies can also use orbital data to forecast regional food supply risks and allocate resources proactively. Space-based monitoring ensures that agriculture remains resilient in the face of climate variability. Advanced orbital technologies bridge observation and action to support sustainable food production.

Forestry management is enhanced by satellites monitoring deforestation, wildfires, and ecosystem health. High-resolution imagery detects illegal logging, tracks wildfire progression, and maps changes in vegetation cover. AI-driven predictive models forecast fire risks, disease outbreaks, and habitat degradation, allowing preventive measures to be implemented. By integrating space-based data into forestry programs, governments and conservation agencies can protect ecosystems, preserve biodiversity, and maintain carbon sequestration. Satellites provide transparent, verifiable information that informs policy, enforces regulations, and enables international collaboration. Space technology allows proactive management of forests to mitigate climate risks while promoting sustainable resource use. Orbital observation transforms forest management into a precise, data-driven process.

Urban infrastructure is increasingly monitored using orbital technologies to mitigate climate risks. Satellites track temperature patterns, flooding, and heat island effects in cities, providing actionable insights for urban planners. AI-driven models analyze the vulnerability of roads, bridges, and buildings to extreme weather and environmental hazards. By integrating satellite data into urban planning, municipalities can design resilient infrastructure, optimize emergency response, and minimize disaster-related economic losses. Satellite imagery also informs long-term adaptation strategies for population growth, climate change, and environmental challenges. Space technology ensures that urban environments are prepared to withstand and adapt to climate emergencies. Orbital monitoring provides cities with the intelligence needed for sustainable, resilient development.

Hydrological systems benefit from continuous space-based monitoring. Satellites measure river flows, groundwater levels, and rainfall distribution, providing critical data for water resource management. AI models integrate this information to predict floods, droughts, and water scarcity, enabling proactive interventions for agriculture, industry, and public consumption. By combining orbital observation with local data, governments can implement adaptive water management strategies that preserve ecosystems and support human needs. Space technology ensures that water distribution is efficient, resilient, and climate-smart. Satellites transform water resource planning into a precise, predictive, and sustainable system. Advanced orbital technologies are indispensable for protecting life, infrastructure, and ecosystems from climate-related water risks.

Climate modeling is strengthened by satellite observations and advanced orbital technologies. Satellites provide continuous data on temperature, cloud cover, sea surface temperatures, and greenhouse gas concentrations, feeding AI-driven climate models. These models simulate potential future scenarios, helping governments, researchers, and industries anticipate climate risks and plan mitigation strategies. By combining observational data with predictive modeling, policymakers can design effective adaptation and mitigation policies. Space-based monitoring ensures that climate models are accurate, timely, and grounded in real-world data. Satellites transform climate science from theoretical projections to actionable intelligence, enabling proactive decision-making.

Early warning systems for extreme weather rely heavily on orbital technologies. Satellites track storm formation, river levels, and wildfire spread in real time, providing alerts to vulnerable regions. AI analyzes satellite data to predict disaster severity, timing, and impact on populations and infrastructure. Governments and disaster management agencies use these warnings to prepare emergency responses, coordinate evacuations, and deploy resources efficiently. Space-based early warning systems reduce human casualties, economic losses, and infrastructure damage during climate events. Satellites provide a constant, impartial, and comprehensive view of threats. Orbital technologies ensure that early warnings are precise, reliable, and actionable.

Policy-making for climate risk mitigation is enhanced through space-based data. Governments use satellite observations to monitor compliance with environmental regulations, track emissions, and assess the effectiveness of mitigation strategies. AI-driven analysis enables policymakers to identify priority areas, allocate resources efficiently, and plan interventions based on evidence. By integrating orbital data into national and international climate policies, decision-makers ensure that adaptation and mitigation efforts are scientifically sound. Satellites provide transparency, accountability, and credibility for climate-related governance. Space technology transforms policy from reactive measures to strategic, proactive planning. Data-driven insights from orbit support sustainable development and climate resilience.

Global collaboration is facilitated through shared space-based climate intelligence. Programs like Copernicus, the UN's SPIDER initiative, and international satellite data sharing agreements allow countries to access high-quality climate information. AI systems integrate data from multiple satellites, regions, and disciplines, providing a unified view of global climate risks. Collaborative monitoring improves early warning systems, disaster preparedness, and climate adaptation planning worldwide. By sharing orbital resources and expertise, nations can address climate challenges collectively, ensuring that even resource-limited countries benefit from advanced monitoring. Space technology becomes a platform for international cooperation, fostering equitable and coordinated climate action.

Monitoring the effectiveness of mitigation and adaptation strategies is strengthened through continuous orbital observation. Satellites track changes in land use, forest cover, glacier retreat, and ocean health, providing feedback on the success of climate interventions. AI-driven analysis identifies areas where strategies are working, and highlights regions needing additional attention. By integrating space-derived data into evaluation processes, governments and organizations can refine policies and programs dynamically. Satellite monitoring ensures that climate mitigation and adaptation efforts are measurable, accountable, and evidence-based. Space technology allows for iterative improvement of strategies, maximizing their impact. Orbital systems provide the intelligence required to achieve long-term climate resilience.

The overall impact of advanced orbital technologies on mitigating climate risks is transformative, providing unprecedented insights and predictive capabilities to safeguard Earth and its inhabitants. Satellites monitor environmental conditions in real time, track greenhouse gas emissions, and observe changes in land, water, and ice systems. AI-driven analytics convert this massive stream of data into actionable insights, enabling governments, industries, and communities to anticipate and respond to climate hazards effectively. By combining observation, prediction, and early warning systems, orbital technologies reduce human, economic, and ecological losses caused by extreme weather and environmental changes. Space-based monitoring ensures that climate adaptation and mitigation efforts are proactive, informed, and precise. Advanced orbital technologies provide a foundation for building resilience in an era of escalating climate risk.

Global agriculture benefits from satellites by adapting practices to changing climate patterns. Crops can be monitored for drought stress, soil moisture levels, and temperature fluctuations, allowing farmers to adjust planting and irrigation strategies. By integrating satellite data with predictive models, food production remains resilient despite extreme weather events. Governments and international organizations use this information to plan food distribution, support vulnerable populations, and maintain regional stability. Satellites transform agriculture into a climate-smart, adaptive system. Space technology ensures that food security is maintained even under increasing environmental pressures.

Water and energy systems gain resilience through space-based climate monitoring. Satellites track river flows, groundwater levels, solar radiation, and wind patterns, informing efficient water management and renewable energy deployment. AI-driven predictive analytics anticipate periods of scarcity or excess, allowing preemptive action to avoid crises. By integrating orbital insights into urban planning, energy grids, and water infrastructure, nations reduce vulnerability to climate extremes. Space technology ensures that essential systems are robust, sustainable, and responsive to environmental changes. Orbital monitoring transforms the management of natural resources into a predictive and adaptive practice.

International collaboration amplifies the benefits of orbital technologies for climate risk mitigation. Countries share satellite data, AI models, and predictive insights to strengthen global early warning systems, disaster response, and policy-making. Collaborative use of space technology ensures that even resource-limited nations have access to critical climate intelligence. By fostering transparency, cooperation, and knowledge sharing, orbital monitoring becomes a tool for global resilience. Satellites provide a neutral platform for collaboration, ensuring equitable access to climate data and solutions. Space technology demonstrates that coordinated action enhances the ability of humanity to address complex environmental challenges collectively.

Ultimately, advanced orbital technologies offer a pathway to a safer, more resilient, and sustainable planet. Satellites, AI-driven analytics, and predictive models enable proactive climate risk management, preserving ecosystems, protecting human lives, and supporting economic stability. By integrating space-based observation into global strategies, nations can mitigate the impacts of climate change while promoting sustainable development. Space technology ensures that humanity is equipped with the knowledge and tools to respond to environmental challenges effectively. The result is a future where climate risks are anticipated, addressed, and minimized through informed, coordinated, and technologically empowered action.

Chapter . 34

Integrating Space Technology into Urban Planning

Designing smarter, safer, and more resilient cities using satellite insights.

Space technology is transforming the way cities are designed, monitored, and managed, providing urban planners with high-resolution data and actionable insights. Satellites offer real-time imagery, environmental monitoring, and spatial analytics that guide infrastructure development, zoning, and transportation planning. AI-driven models interpret these data to optimize urban layouts, traffic flow, and energy consumption. By integrating space technology into urban planning, cities can become more resilient to natural disasters, climate change, and population growth. Satellite monitoring supports sustainable development while enhancing the safety, efficiency, and livability of urban spaces. The combination of orbital observation and AI enables cities to plan intelligently for the challenges of the 21st century.

Traffic and transportation management benefit significantly from space-based monitoring. Satellites track vehicle movement, congestion patterns, and public transportation usage across cities. AI systems analyze this data to optimize traffic signals, public transit routes, and emergency response times. By leveraging satellite insights, urban planners can reduce congestion, lower emissions, and improve overall mobility. Real-time observation also aids in incident detection and rapid intervention, preventing bottlenecks and accidents. Space technology ensures that transportation systems are both efficient and environmentally responsible. Integrating orbital data into transportation planning creates cities that move smoothly and sustainably.

Urban environmental monitoring is enhanced through satellites, which track air quality, temperature, green space distribution, and pollution sources. AI-driven analytics identify areas at risk of heat islands, smog accumulation, or flooding. By using space-derived data, city planners can design mitigation strategies such as planting trees, creating green roofs, and implementing clean energy solutions. Satellite insights also guide zoning regulations and building codes to improve environmental resilience. Space technology allows cities to anticipate and respond to environmental challenges proactively. The integration of orbital monitoring ensures healthier, safer, and more sustainable urban environments for residents.

Disaster preparedness and response are strengthened by integrating space technology into city planning. Satellites monitor flood-prone areas, landslide zones, and seismic activity, providing early warnings to authorities and residents. AI models simulate disaster scenarios, helping urban planners design evacuation routes, emergency shelters, and resilient infrastructure. By combining orbital data with ground-based monitoring, cities can reduce casualties, property damage, and economic losses during natural disasters. Space technology provides a comprehensive perspective that enhances urban safety and resilience. Incorporating satellite insights into emergency planning allows cities to respond quickly and effectively to unforeseen events.

Infrastructure management is optimized through space-based monitoring. Satellites track the condition of roads, bridges, water supply systems, and energy networks, detecting wear, damage, and inefficiencies. AI-driven predictive maintenance schedules repairs before failures occur, extending the lifespan of infrastructure and reducing costs. By integrating orbital data, cities can plan construction, maintenance, and upgrades more effectively. Space technology ensures that urban infrastructure is safe, reliable, and resilient to climate change and population growth. Urban planning informed by satellites allows cities to thrive sustainably while minimizing risks and operational inefficiencies.

Case studies highlight the practical application of space technology in creating smarter and more resilient cities. In Singapore, satellites monitor traffic, air quality, and urban heat islands, providing real-time data to optimize transportation and environmental policies. AI models process this information to adjust traffic light timing, recommend public transit improvements, and identify areas needing green space expansion. By integrating orbital insights, city planners reduce congestion, pollution, and energy consumption. Satellite monitoring also supports emergency preparedness by mapping flood-prone zones and predicting storm impact. Space technology transforms urban management into a data-driven, adaptive, and proactive process.

Smart city initiatives increasingly rely on orbital data to enhance energy efficiency and sustainability. In cities like Barcelona, satellites track solar radiation, building energy use, and electricity consumption across neighborhoods. AI-driven systems analyze these data to optimize heating, cooling, and lighting in buildings while reducing carbon emissions. By integrating satellite insights, urban energy networks become more resilient and efficient, supporting sustainable growth. Real-time monitoring allows for rapid response to energy demand fluctuations or failures. Space technology ensures that urban energy systems are both reliable and environmentally responsible. Satellites empower cities to balance energy needs with climate objectives.

Urban planning for public health benefits from space-based monitoring. Satellites track pollution sources, water quality, and environmental hazards, providing data that inform zoning, green infrastructure, and sanitation policies. AI systems analyze these trends to recommend interventions that reduce health risks, such as planting trees in heat-prone areas or regulating industrial emissions. By incorporating satellite insights, cities can improve air and water quality, enhance recreational spaces, and reduce disease incidence. Space technology enables evidence-based decision-making that prioritizes citizen health. Orbital monitoring transforms urban planning into a proactive strategy for creating safer and healthier communities.

Disaster risk management is strengthened through satellites monitoring earthquakes, floods, and landslides. Cities like Tokyo use space-based observations to update evacuation routes, reinforce critical infrastructure, and plan emergency response. AI-driven predictive models simulate potential disasters and identify vulnerable populations, guiding targeted interventions. By integrating orbital data, urban planners can reduce casualties, property damage, and service disruptions during emergencies. Space technology provides a comprehensive and objective view of risks, supporting resilience planning at the city level. Satellites allow cities to anticipate threats and act before crises occur, ensuring safety and continuity.

Urban expansion and land-use planning are improved through high-resolution satellite imagery. Cities like Dubai and New York analyze growth patterns, monitor construction projects, and assess environmental impacts using orbital data. AI-driven models predict urban sprawl, traffic impact, and resource demand, helping planners make informed decisions about zoning, transportation, and green space allocation. By integrating satellite insights, urban development becomes more sustainable, reducing strain on infrastructure and natural resources. Space technology ensures that urban expansion is strategic, efficient, and environmentally responsible. Satellites enable long-term planning that balances growth with livability and resilience.

Transportation systems benefit greatly from the integration of space technology into urban planning. Satellites provide real-time traffic data, track vehicle density, and monitor public transportation usage across entire metropolitan areas. AI-driven models analyze these data to optimize traffic signals, route planning, and emergency vehicle deployment. By using orbital insights, city planners can reduce congestion, minimize emissions, and improve commuter safety. Satellite monitoring also assists in planning for autonomous vehicles and future mobility solutions. Space technology ensures that transportation infrastructure evolves efficiently and sustainably, adapting dynamically to urban growth and changing demands.

Climate adaptation strategies are enhanced through satellite observation and analysis. Cities use orbital data to monitor rising temperatures, flooding risks, and storm patterns. AI-driven simulations predict the impact of extreme weather events on urban infrastructure, guiding the design of resilient buildings, drainage systems, and green spaces. By integrating satellite insights, municipalities can develop targeted interventions that reduce vulnerability and increase urban resilience. Space technology allows planners to anticipate environmental challenges before they become crises. Satellite data ensures that climate adaptation measures are precise, timely, and scientifically informed. Urban areas can thrive despite changing climatic conditions through proactive, data-driven planning.

Smart utility management is another key application of space technology. Satellites monitor energy usage, water distribution, and waste management systems, providing real-time data to optimize efficiency. AI-driven analysis predicts demand surges, detects leaks, and identifies maintenance needs across utility networks. By combining orbital monitoring with urban management systems, cities can reduce resource waste, minimize costs, and enhance service reliability. Space-based insights also support renewable energy integration and sustainable water management. Satellites enable urban utilities to operate intelligently, resiliently, and sustainably. Space technology ensures that cities provide consistent services while reducing environmental impact.

Urban governance benefits from the integration of AI and satellite data. Municipal authorities use real-time orbital imagery to monitor construction projects, enforce zoning regulations, and plan public services. AI systems analyze demographic, environmental, and infrastructure data to support evidence-based decision-making. By integrating space technology into governance, cities can respond quickly to emerging challenges, allocate resources effectively, and maintain transparency. Satellite insights also enhance citizen engagement by providing accurate information on urban development and environmental conditions. Space technology strengthens governance by enabling cities to operate with intelligence, efficiency, and accountability. Orbital monitoring transforms urban management into a data-driven and proactive process.

Emergency preparedness in urban centers is strengthened through orbital technologies. Satellites detect risks such as floods, fires, earthquakes, and industrial hazards, providing early warnings to authorities and citizens. AI-driven predictive models simulate disaster scenarios, identifying vulnerable populations and infrastructure for targeted interventions. By integrating satellite data into urban emergency planning, cities can reduce casualties, protect property, and maintain essential services during crises. Space technology provides a comprehensive, real-time perspective that supports strategic decision-making. Satellite monitoring ensures that urban emergency response is fast, coordinated, and effective. Advanced orbital technologies make cities safer, more resilient, and capable of withstanding unforeseen challenges.

Case studies demonstrate the effectiveness of space technology in enhancing urban resilience. In New York City, satellites track flooding risks, heat islands, and air pollution levels to inform infrastructure upgrades and emergency planning. AI-driven systems analyze satellite data to optimize stormwater drainage, green space allocation, and energy distribution. By integrating orbital insights, urban planners can prioritize interventions, reduce environmental impact, and improve public safety. Satellite monitoring ensures that city responses to climate hazards are timely, accurate, and evidence-based. Space technology enables cities to adapt dynamically to environmental and population challenges, creating safer, more livable urban areas.

In Tokyo, orbital data supports earthquake and tsunami preparedness. Satellites provide real-time observation of seismic activity, coastal erosion, and urban infrastructure vulnerability. AI models simulate disaster scenarios, allowing planners to optimize evacuation routes, reinforce buildings, and allocate emergency resources efficiently. By combining satellite insights with ground-based sensors, Tokyo can reduce casualties and property damage during natural disasters. Space-based monitoring allows for continuous assessment of urban resilience and infrastructure integrity. Satellite intelligence ensures that cities can respond effectively to unexpected emergencies, enhancing long-term safety. Space technology transforms disaster management from reactive response to proactive planning.

Environmental monitoring in urban areas benefits significantly from satellites. Cities like Singapore and London use orbital data to track air quality, temperature variations, and pollutant dispersion. AI-driven analytics identify pollution hotspots, monitor industrial emissions, and guide policies for cleaner energy and transportation solutions. By integrating satellite data, urban planners can design healthier, more sustainable environments for residents. Satellites provide an impartial and comprehensive view of environmental conditions across the city. Space technology ensures that urban environmental management is proactive, efficient, and evidence-based. Cities can achieve a balance between development, sustainability, and citizen health through space-assisted monitoring.

Smart city initiatives increasingly rely on satellite insights for planning and operational efficiency. Barcelona, for example, uses satellites to monitor traffic, energy consumption, and urban expansion patterns. AI models optimize public transportation, waste management, and energy distribution based on orbital data. By combining space technology with IoT devices and urban sensors, cities can create responsive systems that adapt to changing demands. Satellites provide the high-level perspective needed to coordinate complex urban networks efficiently. Space technology transforms cities into dynamic, intelligent ecosystems capable of sustainable growth. Integrating orbital data ensures that urban development remains forward-looking and resilient.

Long-term urban planning benefits from integrating space technology into strategic decision-making. Satellites provide continuous observation of land use, population density, and environmental conditions, allowing planners to forecast growth patterns and infrastructure needs. AI-driven simulations guide zoning, transportation networks, and resource allocation to ensure sustainable expansion. By incorporating satellite insights, cities can anticipate future challenges such as climate change, population growth, and resource scarcity. Space-based monitoring ensures that urban strategies are evidence-based, adaptable, and resilient. Orbital technologies provide the intelligence required for cities to thrive sustainably while minimizing risk. Space technology ensures that long-term planning is precise, informed, and capable of supporting future generations.

The overall impact of integrating space technology into urban planning is transformative, creating cities that are smarter, safer, and more sustainable. Satellites provide continuous observation of infrastructure, environmental conditions, population growth, and energy use, offering planners a comprehensive perspective. AI-driven analytics interpret this data to optimize urban layouts, transportation systems, emergency response, and resource distribution. By leveraging orbital insights, cities can proactively address challenges such as traffic congestion, pollution, flooding, and energy inefficiency. Space technology ensures that urban development is resilient, efficient, and environmentally responsible. Integrating satellite monitoring into planning allows cities to evolve intelligently in response to changing conditions.

Resilient infrastructure is a key benefit of space-assisted urban planning. Satellites monitor roads, bridges, water systems, and energy networks, detecting vulnerabilities and predicting maintenance needs. AI systems analyze this data to schedule timely repairs, prevent failures, and extend the lifespan of critical infrastructure. By combining orbital observation with predictive analytics, cities reduce costs, minimize disruptions, and enhance public safety. Space technology ensures that urban systems remain reliable and adaptable to environmental challenges. Satellite monitoring transforms reactive maintenance into proactive infrastructure management. Urban resilience becomes measurable, actionable, and sustainable.

Disaster preparedness is strengthened through satellite observation and modeling. Cities can track flood-prone areas, landslides, heat islands, and seismic activity from orbit. AI-driven predictive models simulate potential disasters and guide emergency planning, evacuation routes, and resource allocation. By integrating satellite data, urban authorities can protect populations, reduce property damage, and maintain essential services during crises. Space-based monitoring ensures real-time awareness of risks, allowing for rapid and informed responses. Space technology transforms urban safety into a proactive, data-driven strategy. Satellite insights make cities capable of responding effectively to both natural and human-made hazards.

Environmental sustainability is enhanced by integrating orbital data into city management. Satellites track air quality, energy consumption, vegetation, and climate conditions, informing policies for clean energy, green infrastructure, and pollution reduction. AI systems analyze these data to identify hotspots, optimize urban ecosystems, and plan sustainable growth. By leveraging space technology, cities can reduce emissions, increase energy efficiency, and improve residents' quality of life. Satellite monitoring provides continuous, objective feedback on the effectiveness of sustainability measures. Space technology ensures that urban development balances growth with environmental stewardship. Cities become models of resilience, efficiency, and ecological responsibility.

Ultimately, integrating space technology into urban planning empowers cities to be resilient, intelligent, and future-ready. Satellites and AI provide the tools necessary for proactive decision-making, risk mitigation, and sustainable growth. By leveraging orbital insights, cities can design infrastructure, manage resources, and respond to emergencies with precision and foresight. Space technology enhances urban livability, public safety, and environmental sustainability. The result is a new era of city planning where data-driven, space-enabled strategies create thriving, adaptive, and resilient urban environments. Satellite-assisted planning ensures that the cities of today are prepared to meet the challenges of tomorrow.

Chapter . 35

Space-Based Renewable Energy as a Solution to Power Shortages

Developing orbital solar stations to deliver sustainable energy to Earth.

Space-based renewable energy presents a revolutionary approach to solving global power shortages by capturing solar energy directly from orbit and transmitting it to Earth. Satellites equipped with solar panels can collect sunlight 24/7 without atmospheric interference, weather limitations, or night-time interruptions. The energy is converted into microwaves or laser beams and transmitted to receiving stations on Earth, where it is converted into electricity for practical use. This method provides a continuous, reliable, and sustainable power source that complements terrestrial renewable energy systems. By harnessing energy in space, humanity can overcome geographical and environmental limitations associated with solar and wind power on Earth. Space-based solar energy offers a transformative solution to global energy scarcity and climate challenges.

The technology behind space-based solar energy involves advanced photovoltaic systems, wireless energy transmission, and orbital power management. Satellites carry high-efficiency solar panels that convert sunlight into electricity with minimal energy loss. Power transmission systems, using either microwave or laser beams, deliver energy safely to ground-based rectenna stations. AI-driven monitoring ensures optimal alignment with the sun, maximum energy collection, and precise targeting of energy beams. By combining orbital engineering with intelligent energy management, space-based systems provide a continuous supply of electricity with minimal environmental impact. Space technology ensures that energy can be generated and delivered reliably regardless of location or weather conditions. This innovation has the potential to revolutionize the global energy landscape.

Space-based renewable energy can address energy shortages in remote and underserved regions. Many communities in developing nations face limited access to reliable electricity due to infrastructure challenges or natural barriers. Satellites transmitting solar power provide a solution that bypasses traditional grids, enabling immediate access to clean energy. By using space-based systems, nations can reduce reliance on fossil fuels, improve quality of life, and stimulate economic development. These orbital systems can also integrate with local renewable energy networks, enhancing resilience and sustainability. Space technology ensures that energy equity is achievable on a global scale. Orbital solar power transforms energy accessibility into a practical, scalable solution for all communities.

Environmental benefits of space-based energy are substantial. By reducing dependence on coal, oil, and natural gas, orbital solar power minimizes greenhouse gas emissions and air pollution. Satellites generate energy without disrupting terrestrial ecosystems or requiring large-scale land use, unlike some conventional renewable energy projects. AI-driven energy optimization ensures that electricity is used efficiently and sustainably, further reducing environmental impact. By harnessing solar power in space, humanity can achieve carbon reduction targets more effectively and accelerate the transition to a sustainable energy future. Space-based solar energy exemplifies how orbital technology can combat climate change while meeting global energy demands. Satellites provide an environmentally responsible pathway for sustainable development.

Challenges remain in the development and deployment of space-based renewable energy systems. Engineering satellites to withstand the harsh conditions of space, ensuring efficient energy transmission, and scaling production to meet global demands require significant innovation. Safety protocols must be implemented to prevent interference with aircraft, satellites, or humans from transmitted energy beams. International collaboration and regulatory frameworks are essential to manage orbital space, coordinate energy delivery, and ensure equitable access. Despite these challenges, ongoing research and pilot projects demonstrate the feasibility of space-based energy as a reliable and transformative solution. Space technology provides a practical approach to addressing Earth's energy shortages while promoting global sustainability. Orbital solar power represents a bold step forward in humanity's pursuit of clean and abundant energy.

Several pilot projects and experimental programs have demonstrated the feasibility of space-based solar energy. In Japan, the Space Solar Power Systems initiative has successfully tested wireless energy transmission from orbit to Earth using microwave beams. The United States and the European Union have conducted similar experiments, validating satellite solar collection, transmission efficiency, and ground reception. AI-driven monitoring systems ensure that satellites maintain optimal orientation toward the sun and adjust energy beams precisely to ground stations. These case studies show that continuous, reliable energy delivery from space is achievable with current or near-future technology. By combining orbital engineering with AI and renewable energy infrastructure, space-based systems provide a viable pathway to solve energy shortages globally. Space technology is transforming theoretical concepts into practical, scalable solutions for Earth's energy needs.

Satellite design is critical for efficient space-based renewable energy systems. Solar panels must maximize sunlight absorption while minimizing weight and material costs. Advanced materials, such as high-efficiency photovoltaic cells and lightweight composites, increase energy yield and durability. Satellites require precise stabilization systems to maintain alignment with the sun and target energy beams accurately. AI algorithms continuously monitor satellite performance, predict maintenance needs, and optimize power collection and transmission. By integrating intelligent design with orbital engineering, satellites can operate reliably in the harsh conditions of space. Space technology ensures that energy production from orbit remains consistent, efficient, and sustainable over long durations.

Energy transmission methods are a key component of orbital solar systems. Microwave beams have proven effective for converting solar energy in space and safely delivering it to Earth-based receiving stations. Laser transmission is an emerging alternative that offers precision targeting and potentially higher energy density. Both methods require advanced safety protocols, including automated beam shutoff, precise alignment, and monitoring of atmospheric interference. AI-driven systems continuously adjust energy delivery to maximize efficiency while ensuring safety for people, aircraft, and satellites. By refining transmission technologies, orbital solar systems can provide a reliable, continuous flow of renewable energy. Space-based energy ensures that electricity reaches Earth safely and efficiently, regardless of location or weather conditions.

Integration with terrestrial energy infrastructure is essential for the practical use of space-based solar power. Ground-based rectenna stations convert microwaves or laser energy into electricity compatible with local grids. AI-driven energy management systems optimize distribution, balance supply with demand, and integrate with other renewable sources such as wind or terrestrial solar. By connecting orbital energy systems to existing infrastructure, cities, rural communities, and industrial zones can benefit from a stable, sustainable electricity supply. Satellites provide a complementary solution to Earth-based renewables, filling gaps in regions where sunlight or wind is intermittent. Space technology ensures that energy delivery is continuous, adaptable, and responsive to consumption patterns. Orbital solar power enhances global energy resilience and sustainability.

Economic and policy considerations are crucial for scaling space-based renewable energy. Large investments are required for satellite construction, launch, and maintenance, as well as for ground infrastructure. Governments and international agencies must collaborate to create regulations governing orbital space, energy transmission rights, and equitable distribution. Public-private partnerships can accelerate innovation, reduce costs, and ensure broad access to energy from orbit. By establishing clear policies and investment frameworks, nations can harness space technology to address power shortages effectively. Space-based renewable energy represents a transformative opportunity to provide clean, abundant electricity while supporting economic growth and environmental sustainability. Satellites can deliver energy security on a global scale when combined with strategic planning and collaboration.

Long-term benefits of space-based renewable energy extend far beyond immediate electricity supply. By providing continuous, reliable power, orbital solar systems can reduce dependence on fossil fuels, mitigating greenhouse gas emissions and climate change. Access to clean energy in remote and underserved regions can stimulate economic growth, improve education and healthcare, and enhance overall quality of life. Satellite-collected energy is scalable, adaptable, and capable of supplementing terrestrial renewable sources during periods of low sunlight or adverse weather. By combining space-based power with Earth-bound energy infrastructure, nations can create resilient, carbon-neutral energy networks. Space technology ensures that energy security, environmental sustainability, and social equity are achievable simultaneously.

Global implementation strategies involve coordinated international efforts and technological standardization. Countries must collaborate on satellite design, orbital slots, transmission protocols, and safety regulations to maximize efficiency and minimize risk. Shared research initiatives accelerate innovation while reducing redundant investment, and multinational projects provide equitable access to energy from orbit. AI-driven monitoring systems facilitate integration across borders, ensuring that power is transmitted safely and efficiently. Space-based energy requires a combination of scientific, engineering, and policy collaboration to achieve global impact. Satellites serve as a unifying tool, providing energy solutions that transcend geographic and political limitations. Orbital solar power embodies a cooperative approach to solving worldwide energy challenges.

Environmental impacts of space-based renewable energy are overwhelmingly positive compared to conventional energy systems. Unlike coal or natural gas, satellites generate electricity without producing air pollution, water contamination, or greenhouse gas emissions. Orbital solar panels occupy space in orbit rather than on land, avoiding deforestation, habitat disruption, and competition with agricultural areas. AI-driven energy management ensures that electricity is used efficiently, reducing overall demand and resource strain. By transitioning to space-based energy, humanity can significantly curb environmental degradation while sustaining economic growth. Space technology demonstrates that innovation can meet energy needs while preserving ecosystems. Satellites offer a pathway to a cleaner, safer, and more sustainable future.

Orbital energy systems complement existing renewable sources on Earth. Solar farms, wind turbines, and hydroelectric plants remain critical components of energy networks, while space-based systems provide continuous backup power and grid stability. AI-driven integration allows for smart energy management, balancing supply and demand across terrestrial and orbital sources. By combining these technologies, cities, regions, and nations can ensure uninterrupted electricity even during extreme weather events or periods of low renewable output. Space-based energy reduces reliance on fossil fuels while enhancing the efficiency and resilience of entire energy systems. Satellites transform global energy networks into adaptable, sustainable, and climate-resilient infrastructures. Orbital solar power provides a reliable bridge between current energy capabilities and a fully renewable future.

Investment in space-based renewable energy drives innovation across multiple sectors. Launch technologies, satellite manufacturing, AI analytics, and energy transmission methods benefit from research and development funded by orbital energy initiatives. These advancements have spillover effects in communications, climate monitoring, and industrial automation. By developing space-based energy solutions, humanity gains the dual benefit of solving energy shortages while advancing technology that can address other global challenges. Space technology creates opportunities for economic growth, scientific progress, and environmental stewardship simultaneously. Satellites offer a platform where energy innovation and technological advancement converge to benefit all of humanity. Orbital renewable energy exemplifies the power of space technology to address pressing challenges on Earth.

Technical challenges remain a key consideration in the deployment of space-based renewable energy systems. Satellites must withstand extreme temperatures, radiation, and micrometeoroid impacts while maintaining optimal solar collection. Energy transmission requires precision to avoid inefficiencies or safety risks, and ground-based receivers must be engineered to convert microwaves or laser beams into electricity reliably. AI-driven monitoring and predictive maintenance help mitigate these challenges, allowing satellites to operate continuously with minimal human intervention. Redundant systems, modular designs, and robust materials enhance reliability and longevity. Space technology ensures that engineering challenges are met proactively, creating safe and sustainable energy systems.

Maintenance strategies for orbital solar stations rely heavily on automation and robotics. AI-controlled systems monitor panel efficiency, alignment, and energy output, signaling repairs or adjustments when needed. Robotic maintenance satellites can perform in-orbit repairs, clean panels, or replace components without the need for crewed missions. By using intelligent systems for monitoring and upkeep, operational costs are reduced, and downtime is minimized. Satellites can also communicate with ground stations to provide real-time performance updates, allowing engineers to make strategic adjustments remotely. Space technology ensures that orbital energy systems remain functional and efficient over long periods. Predictive maintenance and autonomous operations are critical for the sustainability of space-based renewable energy.

Public adoption and perception of space-based energy are crucial for its successful implementation. Communities must trust that energy transmission is safe, reliable, and environmentally friendly. Educational campaigns and transparent monitoring can demonstrate the benefits of orbital solar power, addressing safety concerns regarding microwave or laser energy beams. Policymakers and private companies must collaborate to ensure equitable access to energy and to build public confidence. By integrating space-based energy with local infrastructure and renewable initiatives, communities can directly experience its advantages. Space technology allows citizens to benefit from clean, abundant, and reliable electricity while contributing to climate mitigation. Public engagement is essential to transform technological innovation into societal impact.

Integration into national and regional energy grids is a critical step for space-based solar energy. Satellites transmit power to strategically placed ground stations, which then feed into local and national grids. AI-driven grid management optimizes the flow of electricity, balancing supply and demand and integrating terrestrial renewable sources. This ensures continuous power delivery even during periods of variable sunlight or high energy consumption. By connecting orbital energy with existing infrastructure, nations can reduce blackouts, improve efficiency, and transition toward sustainable energy systems. Space technology provides the precision, reliability, and adaptability needed for seamless integration. Satellite-assisted energy networks transform national power systems into resilient and sustainable platforms.

Scaling space-based renewable energy requires global collaboration and investment. Governments, private companies, and international organizations must coordinate on satellite construction, launch schedules, energy transmission protocols, and safety regulations. Shared technological standards and cooperative agreements reduce redundancy, accelerate innovation, and ensure equitable access. By pooling expertise and resources, nations can expand orbital solar power deployment efficiently and sustainably. Space technology becomes a tool for addressing energy shortages worldwide while fostering international collaboration. Orbital renewable energy demonstrates how scientific innovation, strategic planning, and global cooperation can solve one of humanity's most pressing challenges. Satellites offer a pathway to universal, sustainable, and reliable energy.

The transformative potential of space-based renewable energy lies in its ability to provide continuous, reliable, and sustainable electricity to meet global demands. By capturing sunlight directly from orbit, satellites can generate power without the limitations of weather, night, or geographical location. This consistent energy supply can reduce dependence on fossil fuels, lowering greenhouse gas emissions and combating climate change. Integrating orbital solar power with terrestrial energy networks ensures that cities, industries, and rural communities receive uninterrupted electricity. Space technology enables energy systems to be resilient, adaptive, and environmentally responsible. Satellites offer a bold, forward-looking solution to the persistent challenge of global energy shortages.

Space-based renewable energy strengthens energy security on a national and international scale. Countries can diversify their energy sources, reducing vulnerability to supply disruptions or geopolitical conflicts. Orbital systems provide a stable complement to terrestrial renewables, supporting grid stability and preventing blackouts during peak demand periods. By integrating AI-driven energy management, nations can optimize power distribution and anticipate fluctuations in consumption. Satellites ensure that energy availability is reliable, equitable, and sustainable across regions. Space technology empowers nations to secure long-term, uninterrupted energy while reducing reliance on environmentally harmful fuels.

The environmental benefits of orbital solar power extend to air quality, climate mitigation, and land conservation. Unlike fossil fuel plants, space-based energy produces no emissions or pollution. By generating electricity in orbit, satellites reduce the need for large-scale land use for solar farms or hydropower, preserving natural habitats and agricultural areas. AI-driven optimization ensures that energy is used efficiently, minimizing waste and maximizing sustainability. Space technology provides a path for meeting global energy needs while protecting ecosystems and mitigating environmental damage. Satellites exemplify how innovative approaches can reconcile energy production with ecological stewardship. Orbital renewable energy is a sustainable solution for the planet's future.

Economic and technological impacts are equally profound. Investment in space-based energy stimulates innovation in satellite engineering, robotics, AI, and wireless energy transmission. Research and development in these areas generate technological spin-offs that benefit multiple sectors, including communications, defense, and environmental monitoring. Public-private partnerships and international collaborations accelerate deployment and reduce costs, making orbital energy accessible to a wider population. Space technology transforms the energy sector into a hub of innovation, job creation, and sustainable development. Satellites serve as both a source of electricity and a catalyst for technological progress. Orbital renewable energy integrates economic growth with environmental responsibility.

Ultimately, space-based renewable energy represents a paradigm shift in how humanity generates, distributes, and consumes electricity. Satellites provide a reliable, clean, and globally accessible energy source that addresses shortages, reduces environmental impact, and enhances resilience. By harnessing energy in orbit, nations can achieve climate goals, improve quality of life, and stimulate technological and economic growth. Space technology ensures that energy systems are prepared for the challenges of the future while remaining sustainable and efficient. Orbital solar power demonstrates the profound potential of innovation to transform global energy systems. Satellites offer a visionary solution where technology, sustainability, and human progress converge to secure Earth's energy future.

Chapter . 36

Encouraging Private Sector Innovation for Public Good

Fostering partnerships between commercial space companies and governments to solve global challenges.

Private sector innovation plays a crucial role in advancing space technology while addressing public needs on Earth. Commercial space companies have introduced cost-effective launch systems, advanced satellite technologies, and new approaches to orbital infrastructure. By partnering with governments, NGOs, and research institutions, these companies can apply their innovations to solve pressing global problems such as climate monitoring, disaster response, and renewable energy development. Public-private collaborations allow for faster deployment, greater efficiency, and broader impact compared to government-only projects. Space technology innovation driven by the private sector creates opportunities to address societal challenges at scale. Encouraging private sector participation ensures that breakthroughs in space research benefit humanity as a whole.

Incentive programs can stimulate private sector contributions to public good projects. Governments can provide grants, tax benefits, and contracts that reward companies for developing technologies addressing climate change, natural disasters, and energy shortages. By aligning economic incentives with social outcomes, commercial actors are motivated to innovate responsibly. Partnerships can include joint satellite missions, space-based monitoring initiatives, and technology transfer programs that apply orbital advancements to Earth-based challenges. Public-private collaboration amplifies impact while minimizing duplication of efforts. Space technology becomes a tool for social good when commercial innovation is guided by strategic, mission-driven partnerships. Incentives ensure that business objectives and public welfare align effectively.

Collaboration between commercial space companies and governments accelerates technology transfer for societal benefit. Technologies originally developed for space exploration, such as advanced sensors, robotics, and AI, can be adapted for disaster management, agriculture, energy optimization, and healthcare. Public-private partnerships allow for rapid testing, scaling, and deployment of these solutions on Earth. By sharing expertise, data, and infrastructure, governments and companies create an ecosystem where innovation directly improves public welfare. Space technology ensures that private sector advances have tangible, measurable impact on global challenges. Encouraging collaboration fosters a culture of responsible innovation and problem-solving. Satellites and orbital technologies become tools to maximize both commercial and societal benefits.

Ethical frameworks guide private sector engagement in public good initiatives. Companies must prioritize transparency, equity, and environmental responsibility when deploying space-based technologies. Ethical oversight ensures that innovations do not exacerbate inequalities, harm vulnerable populations, or create unintended consequences. Governments and international organizations can establish standards, regulations, and best practices to govern private sector involvement in global challenges. By integrating ethical considerations into innovation strategies, space technology benefits society while minimizing risk. Satellites, AI, and orbital infrastructure operate most effectively when guided by principles of fairness, sustainability, and shared benefit. Private sector innovation achieves its fullest potential when aligned with ethical stewardship.

Global challenges are increasingly addressed through private sector-led space initiatives. Climate monitoring, disaster early warning systems, renewable energy development, and urban planning solutions all benefit from commercial satellite technology and orbital research. Private companies often have the agility to experiment, iterate, and scale solutions faster than government programs. By fostering collaboration, incentivizing responsible innovation, and establishing regulatory oversight, nations can harness private sector capabilities to solve critical problems. Space technology innovation becomes a powerful mechanism for societal improvement when commercial resources, expertise, and creativity are applied strategically. Satellites and private sector ingenuity together create solutions that improve human welfare and protect the planet.

Case studies demonstrate the effectiveness of public-private partnerships in space technology innovation. In the United States, NASA's Commercial Orbital Transportation Services (COTS) program successfully collaborated with companies like SpaceX and Orbital ATK to develop reliable cargo delivery to the International Space Station. These partnerships accelerated innovation while reducing costs and risk for government agencies. Similarly, the European Space Agency works with private firms on satellite data applications for climate monitoring, agriculture, and urban planning. By combining commercial agility with public oversight, these programs achieve both technological advancement and societal benefit. Space technology innovation thrives when governments and private companies align their goals and resources strategically. Satellites and orbital systems become catalysts for public good when commercial capabilities are applied responsibly.

Innovation accelerators and incubators play a critical role in supporting private sector contributions to public challenges. Organizations such as Starburst, Techstars Space, and the European Space Incubator provide funding, mentorship, and networking opportunities for startups developing space-based technologies. These programs encourage the creation of solutions for climate monitoring, disaster response, renewable energy, and other societal needs. By nurturing early-stage companies, accelerators increase the speed and effectiveness of innovation while promoting collaboration with government agencies and research institutions. Space technology benefits from a thriving ecosystem of private innovators, ensuring continuous advancement for public benefit. Public-private partnerships are strengthened when startups have access to resources, expertise, and market opportunities. Satellites and orbital innovations reach their fullest potential through strategic support of commercial initiatives.

Regulatory frameworks are essential to maximize the public impact of private sector space initiatives. Governments must establish clear rules for orbital operations, satellite launches, data sharing, and environmental protection. By creating predictable and fair regulations, authorities encourage private investment while ensuring that space technologies serve societal goals. International agreements, such as the Outer Space Treaty and cooperative space-monitoring programs, provide additional oversight and guidance. Regulatory alignment between countries fosters collaboration, reduces conflict, and ensures safe and equitable use of space resources. Space technology innovation operates most effectively within transparent and accountable regulatory systems. Satellites and orbital solutions benefit both public and private interests when governance structures balance freedom, responsibility, and public good.

Global impact is amplified when private sector innovation is applied to pressing societal challenges. Commercial satellites provide critical data for climate change adaptation, disaster early warning systems, and food security. AI-driven analysis of space-based observations improves decision-making in agriculture, energy management, and urban planning. By combining private technological expertise with public objectives, countries can accelerate progress toward sustainable development goals. Space technology ensures that innovation addresses universal needs rather than serving only commercial interests. Satellites become tools for collective problem-solving, enhancing resilience, equity, and sustainability on a global scale. Public-private collaboration transforms orbital innovation into tangible solutions for humanity's most urgent challenges.

Incentivizing long-term commitment from the private sector ensures sustained progress and societal benefit. Governments can create reward programs, milestone-based funding, and tax benefits for companies delivering impactful solutions. Collaborative research projects, co-developed satellites, and shared data initiatives provide ongoing opportunities for innovation. By linking commercial success to societal outcomes, private companies are encouraged to align technological development with global challenges. Space technology becomes a platform for solving complex problems when private initiatives are guided by public objectives. Satellites, AI, and orbital infrastructure achieve the greatest benefit when commercial innovation is supported, incentivized, and ethically applied. Public-private partnerships ensure that the future of space technology delivers sustainable progress for all.

Technology transfer from private space companies to public applications amplifies the societal impact of innovation. Systems developed for satellite communications, Earth observation, and orbital robotics can be adapted to monitor climate change, manage natural resources, and improve disaster response. By sharing these advancements with government agencies, NGOs, and local communities, commercial innovations accelerate public problem-solving. AI-driven satellite data analytics, originally designed for commercial purposes, can be repurposed for urban planning, agriculture, and public health. Space technology ensures that private ingenuity benefits society beyond commercial profit. Public-private collaboration transforms breakthroughs in orbital technology into solutions for pressing global challenges. Technology transfer maximizes the return on investment for both commercial and societal stakeholders.

Scalability is critical for private sector solutions to have global impact. Satellite networks, AI-driven data platforms, and renewable energy systems developed by commercial companies must be designed for expansion across regions and nations. Modular satellite architectures and cloud-based analytics platforms allow rapid deployment to multiple markets and environments. By scaling solutions effectively, private sector innovation can address energy shortages, climate monitoring gaps, and disaster preparedness needs worldwide. Space technology facilitates scalable solutions by providing universal coverage, continuous monitoring, and adaptable platforms. Public-private partnerships enable innovation to grow beyond isolated pilot projects into global systems. Satellites and orbital solutions achieve maximum societal benefit when designed for broad applicability and expansion.

Education and workforce development are essential components of private sector-driven space innovation. Training programs, internships, and collaborative research initiatives help cultivate a skilled workforce capable of advancing space technology for public good. Partnerships between commercial companies, universities, and governments foster STEM education, encourage innovation, and provide career pathways in satellite engineering, AI, and orbital systems management. By investing in human capital, the private sector ensures the sustainability and continuous growth of space technology applications. Space technology innovation is most effective when supported by a knowledgeable, adaptable, and creative workforce. Satellites and orbital systems serve as both technological tools and educational catalysts for future generations. Workforce development ensures that society fully benefits from space-based innovations.

Entrepreneurship drives creativity and problem-solving in the private space sector. Startups and small companies often take innovative approaches to orbital technologies, from miniature satellites to advanced AI analytics. By fostering entrepreneurial initiatives, governments and investors can encourage solutions that address climate change, energy security, disaster mitigation, and sustainable development. AI-assisted satellite platforms and orbital data services created by entrepreneurial ventures enhance efficiency and accelerate societal benefits. Space technology becomes a testing ground for creative solutions with real-world applications. Private sector entrepreneurship, supported by incentives and collaboration, transforms innovative ideas into tangible benefits for humanity. Satellites, AI, and orbital infrastructure demonstrate how entrepreneurship can meet public needs while advancing technology.

Global collaboration enhances the effectiveness of private sector innovations for public good. Commercial space companies can share data, joint satellite platforms, and technology standards with international partners. These collaborations enable a coordinated response to global challenges such as climate change, food security, and disaster management. Space technology ensures that private innovation is not limited by borders but contributes to universal solutions. By pooling expertise, resources, and infrastructure, nations and companies together can achieve greater societal impact. Satellites, AI, and orbital systems illustrate the potential of cooperative innovation to benefit all humanity. Public-private partnerships facilitate global solutions that are scalable, effective, and equitable.

Funding models are a critical factor in sustaining private sector innovation for public benefit. Governments, venture capital, and international organizations can provide grants, low-interest loans, and milestone-based funding to incentivize commercial companies to develop technologies with societal impact. Public-private investment structures allow risks and rewards to be shared, reducing financial barriers for companies developing space-based solutions. Collaborative funding encourages startups and established firms to focus on energy, climate monitoring, disaster response, and urban resilience projects. By aligning economic incentives with public objectives, space technology can address global challenges effectively. Satellites, orbital infrastructure, and AI systems achieve maximum impact when supported by sustainable financial models. Strategic funding ensures that technological progress benefits both businesses and society.

Risk management strategies are essential in private sector space initiatives. Satellites, launch vehicles, and orbital infrastructure face technical, operational, and environmental risks that could impede societal impact. Companies must implement robust quality control, redundant systems, and contingency plans to ensure mission success. Collaboration with government agencies and research institutions enhances risk mitigation by sharing expertise, resources, and monitoring systems. AI-driven predictive analytics can anticipate failures, optimize maintenance, and minimize downtime, ensuring continuous service delivery. Space technology ensures that private sector innovations remain reliable and effective for public applications. Managing risks responsibly safeguards both investments and societal benefits. Satellites and orbital systems operate most effectively when risk is systematically addressed.

Ethical innovation guides private sector contributions to public good projects. Companies must prioritize transparency, equity, and environmental responsibility when deploying space-based technologies. Ethical oversight ensures that orbital systems do not exacerbate inequalities, endanger communities, or compromise ecosystems. Governments and international bodies can establish frameworks that define acceptable practices, monitor compliance, and promote responsible use of space technology. By integrating ethical principles into innovation, private companies align commercial success with societal welfare. Space technology benefits humanity most when ethical considerations drive its application. Satellites, AI systems, and orbital platforms achieve meaningful impact when innovation is guided by moral responsibility. Ethical innovation ensures that the public good remains central to technological advancement.

Public accountability strengthens the impact of private sector space innovation. Governments, investors, and citizens must be able to monitor, evaluate, and influence how commercial space technologies are applied to societal challenges. Transparency in operations, data sharing, and decision-making fosters trust and encourages responsible deployment of satellites and orbital systems. Independent audits, reporting mechanisms, and stakeholder engagement ensure that projects meet agreed-upon objectives and ethical standards. Space technology becomes a tool for societal improvement when accountability frameworks are in place. Private sector companies delivering orbital solutions are more effective and socially responsible when monitored and guided by public oversight. Satellites, AI-driven analytics, and orbital infrastructure operate most successfully when public interest is prioritized alongside commercial goals.

Collaboration across sectors maximizes the benefits of private space innovation. By integrating efforts from government agencies, non-profits, academic institutions, and commercial companies, projects gain diverse expertise, resources, and perspectives. Joint initiatives can address climate change, renewable energy, urban planning, and disaster response more efficiently and effectively than isolated efforts. Space technology ensures that private sector ingenuity is leveraged for broad societal impact rather than narrow commercial gain. Public-private partnerships, guided by ethical standards, accountability, and strategic investment, transform orbital technologies into tools for global benefit. Satellites and AI-driven platforms become instruments of progress when innovation is coordinated, responsible, and focused on public welfare. Collaboration ensures that private sector advances truly serve humanity.

The overall impact of private sector innovation on public good is transformative, enabling faster, more efficient, and scalable solutions to global challenges. Commercial space companies contribute advanced satellites, AI-driven analytics, and orbital technologies that support climate monitoring, disaster management, renewable energy, and sustainable urban planning. Public-private partnerships accelerate the deployment of these innovations, ensuring that they reach the communities and regions that need them most. By combining the agility of private companies with the oversight, guidance, and funding of governments and international organizations, space technology achieves societal benefits at a scale impossible for a single entity. Satellites and orbital systems become tools for collective problem-solving, resilience, and sustainable development. Private sector innovation ensures that space technology does not remain confined to exploration but delivers tangible improvements on Earth.

Long-term benefits of private sector participation include increased technological capacity, economic growth, and global resilience. Startups and established companies driving space innovation create new industries, workforce opportunities, and technological spin-offs applicable to multiple sectors. By investing in AI, robotics, and orbital systems, these companies indirectly support healthcare, agriculture, energy, and infrastructure solutions. Space technology, when guided by public objectives, generates widespread societal impact while fostering economic development and scientific advancement. Satellites and orbital platforms provide data, insights, and capabilities that strengthen humanity's ability to respond to environmental, social, and economic challenges. Encouraging private innovation ensures that space-based solutions continue to evolve and scale for the greater good.

Societal transformation is accelerated when commercial space technologies are applied to urgent public challenges. Access to orbital data and advanced analytics improves decision-making, resource allocation, and strategic planning for governments, NGOs, and communities. Space technology enhances disaster preparedness, climate adaptation, energy security, and urban resilience. By leveraging private sector ingenuity, humanity gains the tools to address complex, interconnected problems efficiently and effectively. Satellites and orbital systems become agents of change, enabling a future where innovation drives tangible improvements in quality of life, sustainability, and global equity. Public-private collaboration ensures that technological progress is harnessed for the benefit of all.

The future of space-enabled solutions depends on continued partnership, ethical innovation, and strategic investment. Governments and private companies must maintain open communication, shared goals, and accountability frameworks to ensure sustained impact. AI-driven satellite systems, renewable energy platforms, and space-based monitoring technologies will continue to evolve, offering new opportunities for public benefit. By aligning commercial ambition with societal needs, space technology becomes a force for positive transformation, addressing energy shortages, environmental challenges, and global inequality. Satellites, robotics, and AI-driven platforms demonstrate the potential for space-based innovation to provide solutions for Earth's most pressing problems. Private sector engagement ensures that these solutions are sustainable, scalable, and inclusive.

Ultimately, encouraging private sector innovation for public good creates a framework where space technology directly improves life on Earth. Satellites, orbital infrastructure, and AI systems provide critical tools for sustainable development, disaster mitigation, energy security, and environmental stewardship. Public-private partnerships, ethical oversight, and strategic funding ensure that commercial innovation aligns with societal priorities. Space technology becomes a shared resource for humanity, delivering practical benefits, technological progress, and long-term resilience. By fostering collaboration between commercial companies and governments, the future of space-enabled solutions promises a world where innovation and public good advance hand in hand. Satellites offer not only exploration but tangible, positive change for people and the planet.

Chapter . 37

Ethical Considerations of Space Technology on Earth

Ensuring that advancements benefit humanity equitably and do not harm vulnerable populations.

The rapid development of space technology raises significant ethical questions about its use, distribution, and impact on society. Satellites, AI, and orbital infrastructure provide immense benefits, but without careful oversight, they may exacerbate inequalities, invade privacy, or cause environmental harm. Ethical considerations require policymakers, scientists, and private companies to anticipate potential risks and ensure fair access to technology. Decisions about who controls satellite data, how energy systems are deployed, and which communities benefit from space innovations have profound moral implications. Space technology must be guided by principles that prioritize human welfare, environmental protection, and global equity. By integrating ethics into development and deployment, satellites and orbital systems can serve humanity responsibly and sustainably.

Privacy is a primary ethical concern in the use of space technology. Satellites can collect detailed imagery, monitor movements, and gather data on populations, sometimes without consent or transparency. Misuse of this information can infringe on personal freedoms, national sovereignty, and human rights. Ethical guidelines should establish limits on data collection, storage, and sharing, ensuring that monitoring is conducted responsibly. AI-driven analysis of satellite data must also comply with privacy standards, avoiding biased or discriminatory outcomes. Space technology offers unprecedented insight into human activity, but these capabilities must be balanced with respect for individual and collective rights. Satellites should enhance security, public safety, and sustainability without compromising personal freedoms or ethical principles.

Equity in access to space technology is another critical ethical consideration. Advanced satellites, energy systems, and orbital infrastructure are often concentrated in wealthier nations or controlled by large corporations. This can create disparities in who benefits from space innovations, leaving developing countries or marginalized communities behind. Ethical frameworks encourage sharing knowledge, providing affordable access, and supporting capacity-building initiatives worldwide. International collaborations, open-source satellite data, and cooperative space programs can help bridge these gaps. Space technology should serve as a tool for global development and social good rather than an exclusive resource. By promoting equity, satellites and orbital innovations contribute to fairness, opportunity, and shared progress across all regions.

Environmental stewardship is central to the ethical use of space technology. Launching satellites, managing orbital debris, and operating energy systems in space can impact ecosystems both in orbit and on Earth. Debris poses collision risks, while energy transmissions and materials used in satellites must minimize environmental harm. Ethical guidelines call for sustainable design, responsible disposal, and careful monitoring to protect both space and terrestrial environments. AI-driven tracking and management of satellites can reduce collisions, prevent space junk accumulation, and ensure long-term sustainability. Space technology must be developed with a commitment to protecting natural systems and minimizing ecological footprints. Ethical use of satellites integrates environmental responsibility into every stage of design, launch, and operation.

Transparency and accountability are essential for ethical governance of space technology. Governments, private companies, and international organizations must openly communicate objectives, risks, and outcomes of orbital projects. Public reporting, stakeholder engagement, and independent oversight ensure that space innovations align with societal values. Ethical decision-making requires clear responsibilities, legal frameworks, and mechanisms to address potential harm or misuse. Satellites, AI, and orbital infrastructure can deliver enormous benefits, but only if their deployment is transparent and accountable. Space technology must be a force for societal improvement, not unchecked power or exploitation. By prioritizing ethics, humanity ensures that the exploration and use of space enhance equity, sustainability, and public welfare.

Dual-use technology presents a significant ethical challenge in space applications. Satellites and orbital systems designed for civilian purposes, such as climate monitoring or communications, can also be adapted for military or surveillance activities. This dual potential raises concerns about security, privacy, and the militarization of space. Ethical frameworks must ensure that space technology prioritizes peaceful applications and minimizes harm to populations. International agreements and transparency in technology deployment are essential to prevent misuse. Space technology can serve humanity when carefully regulated to balance innovation with moral responsibility. Ethical considerations guide decisions about who benefits, who controls, and how risks are mitigated in dual-use scenarios.

International collaboration is crucial for addressing ethical concerns in space technology. Space is a shared environment, and the actions of one nation or company can affect all others. Cooperative agreements for satellite deployment, debris management, and data sharing promote equitable access and reduce conflict. Ethical collaboration ensures that technological advances serve global needs, including disaster response, environmental monitoring, and sustainable development. Satellites and orbital systems become tools of diplomacy, trust, and shared responsibility when nations work together. Space technology should foster cooperation rather than competition or exclusion. By prioritizing ethical partnerships, humanity ensures that space innovations benefit all countries and communities.

AI integration in space technology introduces additional ethical considerations. Autonomous satellite systems, orbital robotics, and predictive analytics can optimize operations, but they also carry risks of bias, error, and unintended consequences. AI must be designed transparently, with safeguards to prevent discrimination, inequity, or harm. Continuous monitoring, testing, and validation are essential to maintain trust and accountability. Ethical AI ensures that space-based systems for energy distribution, disaster prediction, or resource management operate reliably and fairly. Satellites equipped with AI can improve efficiency and global well-being, but only if algorithms are guided by ethical standards. Space technology becomes a responsible force when AI enhances societal benefits without compromising human values.

Disaster management demonstrates how ethical space technology can serve public good. Satellites provide critical data for early warning systems, relief logistics, and resource allocation during crises. Ethical considerations include ensuring that data reaches vulnerable populations, is not monopolized by powerful entities, and supports equitable response efforts. AI-driven analysis must prioritize human life, safety, and fairness in decision-making. Space technology empowers governments and organizations to act swiftly and responsibly, reducing harm and enhancing resilience. Ethical deployment ensures that satellites are used for protection and humanitarian purposes rather than exploitation or profit. By embedding morality into disaster management systems, space technology maximizes societal benefit.

Balancing innovation with moral responsibility is central to ethical space technology. Rapid technological advancements can outpace regulations, public awareness, and social adaptation. Companies, governments, and international bodies must anticipate risks, establish ethical guidelines, and monitor impacts continuously. Decisions about satellite launches, orbital energy systems, and AI-driven analysis must consider privacy, equity, environmental sustainability, and public welfare. Space technology offers unparalleled opportunities to solve global challenges, but ethical foresight ensures that these solutions are safe, fair, and inclusive.

Satellites, orbital infrastructure, and AI systems achieve their greatest potential when guided by principles of responsibility, justice, and sustainability. Ethical governance transforms innovation from a tool of power into a force for collective good.

Space debris management is a critical ethical concern for the sustainability of orbital operations. Satellites, rockets, and other orbital equipment can leave behind fragments that pose collision risks to functional satellites and spacecraft. These collisions can create more debris, threatening long-term access to space for all nations. Ethical considerations require companies and governments to design satellites for end-of-life deorbiting, minimize debris generation, and implement active debris removal technologies. International cooperation is necessary to establish standards and accountability for debris mitigation. Space technology must be developed responsibly to ensure that innovation today does not compromise future generations' ability to use and benefit from space. Satellites become sustainable tools only when ethical guidelines for debris management are strictly followed.

Climate ethics play an essential role in the deployment of space-based technologies. Satellites and orbital infrastructure can contribute to greenhouse gas monitoring, renewable energy systems, and climate adaptation strategies. However, the environmental costs of launches, materials, and energy-intensive operations must be weighed against the benefits. Ethical frameworks encourage the use of sustainable materials, energy-efficient launches, and environmentally friendly designs. AI-driven monitoring can optimize operational efficiency while minimizing carbon footprints. Space technology demonstrates the potential to protect Earth's environment, but ethical responsibility requires balancing innovation with ecological preservation. Satellites and orbital solutions must support climate goals without creating additional environmental harm.

Energy equity is another vital ethical consideration in space-based solutions. Satellites providing solar energy, communications, or data services have the potential to benefit underserved and remote regions. However, if controlled solely by wealthy nations or corporations, these resources could exacerbate inequality. Ethical governance ensures fair access, affordable deployment, and shared benefits for all communities. Public-private partnerships, international agreements, and regulatory frameworks can help guarantee that space-based energy reaches those who need it most. Space technology becomes a tool for global equity when energy and data access are distributed responsibly. Satellites can bridge gaps in development, education, and sustainability if guided by moral and ethical principles.

Responsible innovation in private sector initiatives is critical to ensuring space technology serves humanity ethically. Companies must prioritize safety, fairness, and long-term impact alongside profit. Transparent operations, independent oversight, and adherence to ethical guidelines help prevent misuse, exploitation, or environmental damage. Private sector satellites and AI systems are most beneficial when integrated into public good projects such as disaster management, climate monitoring, and energy access. Space technology requires a careful balance between commercial innovation and societal responsibility. Ethical stewardship ensures that private innovations in orbit improve quality of life rather than create new risks. Satellites, orbital infrastructure, and AI-driven platforms achieve maximum impact when guided by ethical corporate responsibility.

Long-term societal considerations are central to the ethical use of space technology. Decisions made today regarding satellite deployment, orbital energy systems, AI integration, and space governance will affect future generations. Equity, sustainability, and transparency must be embedded into every stage of technological development. Space debris, climate impact, and resource allocation must be managed with foresight and care. Ethical principles ensure that the benefits of space technology are durable, accessible, and socially just. Satellites become instruments of long-term societal progress when innovation is guided by responsibility, collaboration, and fairness. By embedding ethics into every decision, humanity can use space technology to build a safer, more equitable, and sustainable world.

Responsible data sharing is a cornerstone of ethical space technology. Satellites collect vast amounts of information, including environmental, agricultural, and population data, which can support research, disaster response, and policy decisions. However, sharing this data must be done transparently, securely, and with respect for privacy and national sovereignty. Ethical frameworks encourage open access for public benefit while protecting sensitive information from misuse. Public-private collaborations, international agreements, and standardized protocols can ensure that satellite data contributes to global welfare rather than concentrating power or knowledge in the hands of a few. Space technology becomes a tool for societal progress when data is shared responsibly, fostering transparency, collaboration, and equity. Satellites provide insights that benefit humanity only when ethical data governance guides their use.

AI governance is critical in space-based systems to prevent unintended harm. Autonomous satellites, orbital robotics, and predictive analytics can optimize operations but also carry risks of bias, errors, or misuse. Ethical AI ensures that algorithms are transparent, auditable, and designed to prioritize human welfare. Continuous monitoring, testing, and validation prevent discriminatory outcomes, privacy violations, or unintended ecological consequences. Space technology integrates AI responsibly by establishing clear accountability, oversight mechanisms, and safety protocols. Satellites equipped with AI can improve energy distribution, disaster response, and climate monitoring, but only under ethical governance. Ethical AI ensures that orbital innovations advance societal benefits without compromising rights, safety, or equity.

International treaties and agreements provide a framework for ethical governance in space. The Outer Space Treaty and other multilateral agreements establish principles of peaceful use, shared access, and environmental protection. Compliance with these treaties ensures that satellites, orbital infrastructure, and space-based energy systems benefit humanity collectively. Ethical international collaboration mitigates conflict, fosters trust, and promotes equitable distribution of technology and resources. Space technology operates most effectively when nations cooperate under shared moral and legal standards. Satellites and orbital systems demonstrate global responsibility when guided by treaties that prioritize public good over national or commercial advantage. Ethical frameworks at the international level ensure fairness, security, and sustainability.

Humanitarian applications highlight the societal value of ethical space technology. Satellites provide critical support for disaster relief, famine prevention, and emergency communication. Ethical use ensures that orbital systems prioritize vulnerable populations, equitable aid distribution, and transparency in decision-making. AI-driven analytics combined with satellite data can improve humanitarian response efficiency and effectiveness. Space technology can save lives, reduce suffering, and protect communities when deployed responsibly. Ethical considerations prevent exploitation, data monopolization, or harm during humanitarian operations. Satellites serve as instruments of compassion and justice when guided by principles of responsibility, fairness, and equity.

Ethical education in space technology is essential for cultivating responsible innovation. Scientists, engineers, policymakers, and private companies must understand the moral implications of their work. Educational programs, professional guidelines, and ethics training promote awareness of privacy, equity, sustainability, and long-term societal impact. By integrating ethical principles into curricula and workplace practices, the next generation of space professionals can make informed, responsible decisions. Satellites, orbital systems, and AI platforms operate most effectively when developers are guided by strong ethical foundations. Space technology becomes a tool for global good when innovation is paired with moral responsibility and continuous ethical learning. Education ensures that ethical considerations remain central to all space endeavors.

Ethical principles are essential to ensure that space technology benefits all of humanity. Satellites, orbital infrastructure, and AI-driven systems can solve major global challenges, but only when development and deployment are guided by fairness, sustainability, and transparency. Long-term governance frameworks help regulate access, manage resources, and prevent misuse, ensuring that space remains a shared domain for peaceful and equitable purposes. Ethical oversight encourages companies and governments to prioritize societal welfare, environmental stewardship, and global equity in their projects. Space technology achieves its full potential when innovation aligns with moral responsibility and public interest. By embedding ethics into decision-making, humanity can harness orbital systems to improve quality of life while safeguarding future generations.

Sustainability in space operations reinforces ethical responsibility. Satellites and orbital systems must be designed to minimize debris, reduce environmental impact, and maximize lifespan. Launches, materials, and energy use should follow best practices that preserve both Earth's ecosystems and the orbital environment. Ethical guidelines ensure that technological progress does not come at the cost of environmental degradation. Space technology can advance energy, communication, and monitoring capabilities while remaining ecologically responsible. Satellites exemplify how innovation and sustainability can coexist when guided by ethical considerations. Long-term benefits depend on balancing advancement with environmental stewardship.

Global responsibility extends to equitable access and international collaboration. No single nation or company should monopolize orbital resources or data, as this can exacerbate inequality and limit the benefits of space technology. Treaties, agreements, and cooperative initiatives ensure that orbital systems serve collective needs, including disaster management, climate adaptation, and energy equity. Ethical frameworks promote trust, transparency, and shared governance, allowing space innovations to address humanity's greatest challenges collaboratively. Satellites and orbital infrastructure provide critical capabilities that must be managed for the common good. Responsible use of space technology fosters global solidarity, cooperation, and resilience.

Innovation must be paired with accountability to achieve societal benefits. Public-private partnerships, regulatory oversight, and continuous ethical review ensure that satellite systems and AI-driven platforms serve public interest rather than narrow commercial or political goals. Transparency in data collection, deployment, and usage builds trust among stakeholders, enabling widespread adoption of space-enabled solutions. Ethical accountability ensures that orbital technologies improve quality of life, protect vulnerable populations, and contribute to sustainable development. Space technology becomes a tool for empowerment, progress, and global well-being when guided by responsibility and oversight. Satellites can drive innovation while serving as instruments of fairness, equity, and justice.

Ultimately, ethical considerations are the foundation for using space technology to benefit Earth. Satellites, orbital energy systems, AI platforms, and communication networks offer transformative solutions, but only if guided by moral principles, transparency, sustainability, and equity. Long-term governance, responsible innovation, and global collaboration ensure that these technologies address humanity's pressing challenges effectively and fairly. Ethical stewardship transforms space technology from a purely scientific or commercial endeavor into a vehicle for public good, social justice, and planetary resilience. By prioritizing ethics in every aspect of development and deployment, satellites and orbital systems can secure a future where space innovation supports humanity, the environment, and global well-being.

Chapter . 38

Preparing for the Next Generation of Space Applications

Anticipating technologies that will revolutionize Earth's industries and ecosystems.

The next generation of space applications promises to transform life on Earth in unprecedented ways. Emerging technologies, such as advanced satellite constellations, space-based AI systems, and orbital manufacturing, are poised to address global challenges more efficiently than ever before. Preparing for these innovations requires foresight, strategic planning, and investment in research and infrastructure. Governments, private companies, and academic institutions must collaborate to ensure that new space technologies are developed responsibly and deployed equitably. Satellites and orbital systems will continue to provide solutions for climate monitoring, disaster response, energy management, and sustainable development. By anticipating future capabilities, humanity can harness space technology to maximize societal benefit while mitigating potential risks.

Training and education are fundamental to preparing the next generation of space applications. Developing expertise in orbital mechanics, AI, robotics, satellite communications, and materials science is critical to fully exploit emerging space technologies. Universities, research institutions, and private companies must provide programs that equip students and professionals with the knowledge and skills necessary for advanced space applications. Interdisciplinary education combining engineering, ethics, policy, and environmental science ensures that space technologies are developed responsibly. Satellites and orbital systems require a workforce capable of designing, maintaining, and optimizing these advanced tools. Preparing future talent guarantees that the benefits of space innovation are realized efficiently and safely.

Investing in research and development is key to driving the next wave of space applications. Experimental technologies such as orbital solar power stations, space-based manufacturing, and autonomous satellite swarms have the potential to revolutionize energy, industry, and logistics on Earth. Funding long-term projects, encouraging collaboration between sectors, and supporting high-risk innovations accelerate the deployment of transformative technologies. By prioritizing research, humanity ensures that space applications evolve to meet critical societal needs. Satellites and orbital systems become platforms for continuous improvement, addressing emerging global challenges. Strategic investment today secures the technological foundations of tomorrow's space-enabled solutions.

Regulatory frameworks and governance structures must evolve alongside technological advances. The next generation of space applications will operate at unprecedented scales, involving mega-constellations, orbital energy systems, and international data networks. Policymakers must establish rules for orbital traffic management, spectrum allocation, debris mitigation, and ethical AI use. Clear regulations ensure safety, equitable access, and the prevention of monopolization or misuse. Satellites and orbital platforms can achieve maximum benefit when governance frameworks anticipate emerging technologies and guide their responsible deployment. Preparing for the next generation of space applications requires proactive policy development. Space technology reaches its full potential when innovation is supported by governance that safeguards public good.

Global collaboration is essential for maximizing the impact of new space applications. Emerging technologies will require shared infrastructure, data, and expertise to address issues such as climate change, renewable energy, disaster mitigation, and sustainable urban planning. International cooperation ensures equitable access, reduces redundancy, and promotes transparency in the deployment of space systems. Satellites, AI, and orbital innovations can solve global challenges only when countries, institutions, and companies work together toward common objectives. Preparing for the next generation of space applications requires a shared vision, coordinated investment, and collaborative governance. Space technology becomes a tool for collective progress when global collaboration drives development, deployment, and ethical use.

Emerging technologies are shaping the future of space applications and their impact on Earth. Innovations in miniaturized satellites, high-bandwidth communication systems, and orbital platforms enable unprecedented data collection and global connectivity. Advanced propulsion systems, autonomous spacecraft, and reusable launch vehicles reduce costs and increase access to space for research, commercial, and humanitarian purposes. These technologies allow satellites to monitor climate patterns, optimize agricultural production, and support energy infrastructure more efficiently than ever before. Space-based innovation also includes orbital manufacturing, which can produce materials in microgravity with superior strength and durability. Preparing for the next generation requires understanding, developing, and integrating these technologies responsibly.

AI integration is transforming how space applications operate and deliver benefits to Earth. Autonomous satellite systems can manage constellation coordination, data analysis, and orbital adjustments with minimal human intervention. AI-driven insights from satellite imagery enhance disaster prediction, environmental monitoring, and urban planning, providing actionable intelligence for governments and communities. Machine learning algorithms optimize energy distribution, resource management, and communication networks from orbit. Ethical deployment of AI ensures reliability, fairness, and transparency, avoiding biased or unsafe outcomes. Space technology combined with AI accelerates problem-solving for global challenges, allowing faster and more precise responses. Preparing for future applications requires balancing technological capabilities with ethical oversight.

Advanced robotics are central to next-generation space applications. Robotic satellites, orbital drones, and autonomous repair units enable in-orbit maintenance, debris removal, and construction of space infrastructure. These systems reduce the need for human presence in hazardous conditions, lowering operational risk and costs. Robotics also facilitates modular satellite networks that can be upgraded or reconfigured as technology advances. By automating repetitive and high-risk tasks, robotics maximizes efficiency and ensures reliability in space systems. Ethical and technical considerations must guide robotic deployment to prevent accidents, collisions, and misuse. Satellites and orbital infrastructure become safer, more adaptable, and sustainable when robotics is integrated responsibly. Preparing for the next generation requires designing robotic systems that complement human oversight.

Orbital infrastructure is evolving to support advanced space applications. Mega-constellations, space-based energy stations, and orbital manufacturing platforms create new possibilities for addressing Earth's challenges. Robust orbital networks allow continuous monitoring, rapid data transmission, and high-capacity communication that supports disaster response, agriculture, and urban management. Infrastructure planning must consider collision risks, orbital debris mitigation, and equitable access to ensure long-term sustainability. Satellites operating in coordinated networks maximize efficiency, coverage, and reliability for global benefit. Preparing for next-generation applications requires building orbital infrastructure that is resilient, scalable, and ethically managed. Space technology reaches its full potential when infrastructure supports both innovation and societal impact.

The future of Earth observation systems is being reshaped by advanced satellites, sensors, and AI analysis. Next-generation satellites provide higher-resolution imagery, real-time monitoring, and multispectral data that improve environmental assessment, resource management, and disaster response. Combined with AI, these systems can detect patterns, predict crises, and guide policy decisions with unprecedented accuracy. Ethical access and data sharing ensure that these benefits are available to all nations, supporting equitable development and global sustainability. Preparing for next-generation applications means designing observation systems that are precise, scalable, and accessible. Satellites become instruments for informed decision-making when technological capabilities are combined with responsible governance. Space technology thus provides a foundation for smarter, more resilient, and sustainable societies.

Workforce development is crucial to prepare for the next generation of space applications. Highly skilled professionals in satellite engineering, AI, robotics, materials science, and orbital logistics are needed to design, operate, and maintain advanced space systems. Universities, vocational programs, and private sector initiatives must provide targeted training to build expertise in emerging technologies. Interdisciplinary knowledge combining engineering, ethics, policy, and environmental science ensures that new space applications are developed responsibly and effectively. Satellites and orbital infrastructure rely on a capable workforce to maximize societal impact. Preparing future generations guarantees that technological progress translates into practical solutions for Earth's industries and ecosystems. Workforce readiness ensures that innovation is sustainable, safe, and globally beneficial.

Interdisciplinary collaboration enhances the effectiveness of next-generation space applications. Combining expertise from engineering, computer science, environmental science, economics, and policy creates holistic solutions to complex global challenges. Satellites, orbital infrastructure, and AI systems require collaboration between governments, private companies, and academic institutions to address energy, climate, disaster response, and sustainable development issues. Interdisciplinary teams foster innovation by integrating diverse perspectives, improving problem-solving, and anticipating unintended consequences. Preparing for the next generation of space applications requires building networks that leverage cross-sector expertise. Satellites and orbital systems achieve maximum societal benefit when collaboration bridges technical, ethical, and policy domains.

Integrating multiple disciplines ensures solutions are efficient, equitable, and sustainable.

Innovation hubs are essential for nurturing the next generation of space technology. These hubs provide incubators, accelerators, and research centers that bring together startups, established companies, academic institutions, and investors. By fostering creativity, experimentation, and collaboration, innovation hubs accelerate the development and deployment of transformative space applications. Satellites, AI-driven platforms, and orbital systems benefit from environments where knowledge, technology, and resources are shared. Preparing for future applications requires investing in hubs that cultivate innovation while maintaining ethical standards and societal focus. Space technology thrives in ecosystems that combine technical expertise, strategic funding, and public-private collaboration. Innovation hubs transform ideas into scalable solutions that address real-world challenges.

Funding strategies are vital to support next-generation space applications. Governments, private investors, and international organizations must provide stable, long-term financing for research, development, and deployment of satellites, orbital infrastructure, and AI systems. Public-private partnerships and milestone-based funding encourage innovation while aligning commercial goals with societal benefits. Funding mechanisms should also prioritize projects that promote sustainability, equity, and ethical responsibility. Satellites and space-based systems require consistent investment to maintain technological advancement and maximize impact. Preparing for the next generation of space applications involves balancing risk, reward, and public good through strategic funding. Space technology reaches its full potential when financial resources are paired with long-term vision and ethical oversight.

Integrating emerging space technologies into Earth-focused solutions ensures that innovation delivers tangible benefits. Next-generation satellites, AI, robotics, and orbital infrastructure can enhance energy distribution, disaster response, climate monitoring, and urban planning. By focusing on practical applications, space technologies improve quality of life, economic growth, and environmental sustainability. Collaboration between private companies, governments, and research institutions ensures that solutions are scalable, inclusive, and effective. Preparing for the next generation of space applications requires translating orbital capabilities into actionable outcomes on Earth. Satellites and orbital systems become instruments of progress when innovation is aligned with societal needs. Space technology has the power to revolutionize industries and ecosystems when applied strategically and responsibly.

International standards are essential for coordinating the next generation of space applications. Mega-constellations, orbital energy systems, and advanced satellite networks require globally accepted norms for traffic management, spectrum allocation, and debris mitigation. Harmonized standards prevent collisions, reduce duplication of effort, and promote equitable access to orbital resources. Satellites, AI-driven platforms, and orbital infrastructure operate more efficiently when guided by shared protocols. Preparing for future space applications involves collaboration among nations, agencies, and private companies to ensure interoperability and safety. International agreements enable responsible innovation while maximizing societal and environmental benefits. Global coordination ensures that space technology serves humanity collectively rather than individual interests.

Ethical deployment remains a core principle in next-generation space technology. As satellites, orbital energy stations, and AI systems become more capable, the potential for misuse or unintended harm grows. Companies, governments, and international bodies must prioritize fairness, transparency, and environmental stewardship. Ethical frameworks guide decision-making regarding data access, energy distribution, and planetary resource utilization. Satellites can provide global benefits only when their deployment aligns with moral responsibility and equitable practices. Preparing for advanced space applications requires integrating ethics into design, launch, and operational strategies. Responsible use ensures that technological progress enhances global well-being without creating inequality or risk.

Adaptive regulation is critical to keep pace with rapid advancements in space technology. Regulatory frameworks must be flexible enough to address emerging challenges such as autonomous satellites, orbital manufacturing, and AI-driven systems. Agencies should anticipate potential risks, establish clear guidelines, and periodically update policies as technology evolves. Satellites, orbital platforms, and space-based energy networks require oversight that balances innovation with safety, equity, and sustainability. Preparing for next-generation applications involves proactive policy development that supports growth while minimizing environmental and societal harm. Adaptive regulation ensures that space technology develops responsibly and delivers tangible benefits. By anticipating future scenarios, governance can guide space applications toward long-term global benefit.

Technological forecasting is essential to prepare for the evolution of space applications. Predictive modeling of satellite capabilities, orbital infrastructure, and AI advancements allows governments and companies to plan investments, resource allocation, and deployment strategies. Forecasting identifies potential challenges, including orbital congestion, data security, and environmental impact, allowing mitigation before problems escalate. Satellites, AI systems, and orbital infrastructure benefit from foresight that anticipates technical, ethical, and societal considerations. Preparing for the next generation requires not only building capabilities but also envisioning their potential impacts on Earth. Forecasting guides research priorities, funding, and collaboration to ensure that space technology achieves maximum societal benefit. Space innovation is most effective when informed by careful anticipation and strategic planning.

Strategic planning unites workforce development, funding, regulation, ethics, and technology to prepare for next-generation space applications. Coordinated efforts ensure that satellites, orbital platforms, and AI systems are designed, deployed, and managed for maximum societal impact. Public-private partnerships, international collaboration, and cross-disciplinary teams provide the expertise and resources needed to tackle complex global challenges. Preparing for future applications requires integrating technical, ethical, and policy considerations into long-term strategies. Space technology can address climate change, energy distribution, disaster response, and urban planning when guided by foresight and strategy. Satellites and orbital systems reach their full potential when innovation is systematically aligned with societal goals. Strategic preparation ensures that the next generation of space applications delivers sustainable, equitable, and transformative solutions.

Future trends in space technology promise to reshape industries, ecosystems, and daily life on Earth. Advanced satellite constellations, AI-driven analytics, orbital manufacturing, and space-based energy systems will deliver solutions for climate monitoring, disaster response, sustainable agriculture, and global connectivity. Preparing for these innovations ensures that their deployment maximizes societal benefit while minimizing risks to people and the environment. Strategic planning, interdisciplinary collaboration, and ethical oversight are essential to harness the full potential of next-generation space applications. Satellites and orbital infrastructure will not only explore space but also serve as instruments for improving life on Earth. By anticipating technological capabilities and societal needs, humanity can integrate space solutions effectively and responsibly.

Commercialization plays a central role in accelerating the development and deployment of next-generation space applications. Private companies provide funding, innovation, and operational efficiency, complementing government research and public initiatives. Public-private partnerships allow satellites, AI systems, and orbital platforms to scale quickly and reach underserved regions. Ethical frameworks and regulatory guidance ensure that commercialization does not compromise equity, safety, or environmental sustainability. Preparing for future applications requires balancing profit motives with public interest, ensuring that space technology benefits everyone. Satellites and orbital innovations become tools for global development when commercial innovation is paired with responsibility.

Societal impact is a key measure of success for next-generation space applications. Satellites provide real-time data to improve energy efficiency, resource management, disaster mitigation, and urban planning. AI-driven analysis and orbital infrastructure enhance decision-making, allowing communities to respond quickly to environmental, economic, and social challenges. Preparing for the future means ensuring that these technologies are accessible, inclusive, and designed to address real-world problems. Space technology will influence industries, governance, and everyday life, creating opportunities for sustainable development and improved quality of life. Ethical foresight and strategic planning maximize positive outcomes for both local and global communities.

Ethical oversight ensures that innovation in space technology remains aligned with human and environmental values. Transparency, accountability, and responsible governance prevent misuse, inequality, and harm. Satellites and orbital systems are most effective when their applications are monitored, regulated, and continuously evaluated. Preparing for the next generation of space applications requires embedding ethical principles into every stage of research, development, and deployment. Space technology achieves long-term societal benefit when guided by moral responsibility and global collaboration. Ethical oversight transforms innovation from a mere technical advancement into a tool for collective progress.

The roadmap for integrating next-generation space technology into everyday life relies on education, collaboration, regulation, and sustainable deployment. Satellites, orbital infrastructure, AI systems, and space-based energy platforms will become integral to addressing Earth's most pressing challenges. Preparing for these applications ensures that technology improves quality of life, strengthens resilience, and promotes equity. Strategic vision, international cooperation, and ethical governance guarantee that space innovation remains a force for global good. Satellites and orbital systems, thoughtfully applied, will connect people, monitor our planet, and provide solutions to the crises of today and tomorrow. By preparing today, humanity ensures that the next generation of space applications delivers sustainable, transformative, and equitable benefits for all.

Chapter . 39

The Role of Policy in Maximizing Space Benefits for Earth

Creating regulations and frameworks to ensure space technology contributes positively to society.

Policy plays a fundamental role in shaping how space technology benefits life on Earth. Governments, international bodies, and regulatory agencies develop frameworks that guide the development, deployment, and use of satellites, orbital infrastructure, and space-based systems. Effective policy ensures that technological innovations are safe, equitable, and sustainable, preventing misuse or environmental harm. By establishing clear rules and standards, policymakers can maximize societal benefits from space applications such as disaster management, climate monitoring, energy distribution, and communications. Satellites and orbital platforms are most effective when guided by policy that aligns innovation with the public good. Preparing and implementing thoughtful regulations ensures that space technology serves humanity responsibly and efficiently.

International coordination is essential to ensure that space benefits are shared equitably across nations. Orbital space is a shared environment, and satellites operated by one country can impact the operations of others. International agreements, such as the Outer Space Treaty, provide a foundation for cooperative use, collision avoidance, and sustainable resource management. Collaborative policy frameworks ensure that satellites, AI systems, and orbital infrastructure are used for collective benefit rather than national or commercial advantage. By harmonizing regulations, nations can prevent conflicts, reduce duplication, and promote innovation. Space technology maximizes its impact when guided by policies that encourage cooperation and equitable access. Global collaboration ensures that space applications contribute to sustainable development for all countries.

Funding and incentives are critical aspects of policy for maximizing space benefits. Governments can provide grants, tax incentives, and public-private partnerships to encourage investment in satellite networks, orbital energy systems, and space-based AI applications. Strategic incentives direct private sector innovation toward societal challenges such as renewable energy, disaster mitigation, and climate monitoring. Satellites and orbital platforms achieve greater societal impact when financial policies encourage development and deployment for public good rather than purely commercial gain. Preparing for the future requires policymakers to design funding structures that balance risk, reward, and social benefit. Space technology becomes a force for global improvement when policy aligns economic incentives with public priorities. Funding mechanisms must support both technological advancement and equitable societal outcomes.

Regulatory oversight ensures safety, accountability, and ethical conduct in space applications. Policies govern satellite launches, orbital debris management, spectrum allocation, and AI-driven space systems. Clear regulations reduce risks, protect shared resources, and prevent monopolization or misuse of space technology. By providing oversight, governments ensure that satellites and orbital infrastructure operate reliably and ethically. Policies must also evolve to keep pace with rapid technological advancements, such as autonomous satellites, mega-constellations, and orbital manufacturing. Satellites and space-based AI systems maximize their societal value when supported by proactive, adaptive regulation. Governance frameworks ensure that space innovation contributes to human welfare, environmental protection, and global security.

Public engagement is a vital element of effective space policy. Citizens, communities, and stakeholders must have access to information about satellite data, space-based projects, and policy decisions that affect society. Transparent communication and participatory mechanisms enable people to influence how space technology is deployed for public benefit. Public input ensures that policies reflect societal priorities, including equity, sustainability, and ethical responsibility. Satellites, AI platforms, and orbital systems are most effective when governed in a manner that incorporates public interest. Preparing for next-generation applications requires policies that balance innovation, oversight, and democratic accountability. Space technology achieves maximum benefit when policies are designed with both expertise and public participation.

Policy for innovation plays a crucial role in guiding the development of space technologies that address societal challenges. Governments can establish frameworks that encourage research and development in satellites, orbital infrastructure, AI-driven platforms, and space-based energy systems. Policies that support experimentation, pilot projects, and collaboration between private companies and public institutions accelerate the creation of solutions for climate monitoring, disaster response, and sustainable urban planning. Regulatory incentives can reward projects that prioritize societal benefit, environmental stewardship, and equitable access. Satellites and orbital systems achieve greater impact when innovation is guided by clear, forward-thinking policies. By aligning technological progress with social priorities, space technology can deliver meaningful improvements to life on Earth.

Disaster response is an area where policy maximizes the practical benefits of space technology. Satellites provide real-time data for monitoring hurricanes, floods, wildfires, and earthquakes, enabling faster and more coordinated emergency responses. Policies can ensure that satellite data is openly accessible to governments, NGOs, and communities, while respecting privacy and security. Governments can establish guidelines for integrating space-based information into disaster management frameworks, improving decision-making and resource allocation. Space technology becomes a life-saving tool when policies guarantee timely access and ethical usage. Preparing and enforcing disaster-related regulations ensures that satellites and orbital systems contribute effectively to resilience and safety. By linking policy with technological capability, society benefits from a more responsive and proactive disaster management system.

Climate monitoring is another critical area where policy shapes space technology's impact. Satellites can track greenhouse gases, deforestation, ocean temperatures, and ice cover, providing essential data for climate mitigation and adaptation strategies. Policies that mandate data sharing, standardize measurements, and encourage collaboration improve the accuracy and usefulness of satellite observations. Governments can incentivize private sector contributions to climate monitoring, ensuring that emerging technologies support global sustainability goals. Satellites and orbital systems become instruments of environmental stewardship when policy frameworks enforce accountability, transparency, and responsible use. By integrating climate priorities into space policy, technology contributes directly to reducing emissions, protecting ecosystems, and guiding sustainable development.

Global security depends on policies that balance technological advancement with ethical use. Satellites, AI platforms, and orbital infrastructure have dual-use potential, providing both civilian and defense applications. Clear regulations and international agreements help prevent misuse, militarization, or conflict escalation in space. Policies can also encourage transparency, debris management, and cooperative surveillance to enhance global safety. Satellites maximize societal benefit when their use is governed by international norms, ethical oversight, and cooperative security measures. Preparing for next-generation space applications requires anticipating threats, establishing preventive policies, and ensuring compliance across nations. Policy ensures that technological progress strengthens security without compromising human welfare or equity.

Space commercialization requires careful policy guidance to ensure equitable access and public benefit. Private companies provide innovation, funding, and operational capabilities for satellites, orbital energy, and AI-driven systems. Policies can establish guidelines for licensing, spectrum allocation, data access, and ethical operations, preventing monopolies or misuse. Public-private partnerships and regulatory incentives align commercial interests with societal goals. Satellites and orbital infrastructure contribute maximally to Earth's development when commercialization occurs under policies that emphasize sustainability, transparency, and equity. Preparing for next-generation applications requires proactive regulatory frameworks that foster both innovation and responsible deployment. Space technology reaches its full potential when commercial efforts are guided by sound policy and public oversight.

Data governance is a critical element of space policy that ensures satellite and orbital information is used responsibly. Policies can establish rules for data collection, storage, sharing, and privacy, balancing transparency with protection of sensitive information. AI-driven analysis of satellite data must comply with ethical guidelines to prevent bias, misuse, or inequitable access. Regulatory frameworks can enforce standards for interoperability, quality, and accessibility of space-derived data. Satellites and orbital systems achieve maximum societal benefit when their information is governed by robust, transparent, and ethical policies. Preparing for next-generation space applications requires policies that safeguard both public trust and technological utility. Proper data governance ensures that space technology contributes to informed decision-making globally.

Equitable access to space technology is another crucial policy focus. Satellites, AI platforms, and orbital infrastructure should be accessible to all nations, communities, and sectors, especially those that are underserved or vulnerable. Policies can promote shared infrastructure, affordable access, and capacity-building initiatives in developing regions. By ensuring fairness in the deployment of space technology, governments and international bodies prevent disparities that could exacerbate global inequality. Satellites serve humanity most effectively when equitable access is prioritized alongside technological advancement. Preparing for the future requires policy mechanisms that remove barriers and enable all countries to benefit from space innovations. Equity in access guarantees that the transformative potential of space technology is shared by everyone.

International cooperation is central to effective space policy. Space is a shared domain, and satellite operations often affect multiple countries. International agreements, treaties, and cooperative initiatives ensure that satellites and orbital systems are managed safely, ethically, and sustainably. Collaborative policy frameworks promote joint monitoring, disaster response, climate tracking, and orbital debris management. By working together, nations maximize the benefits of space technology while minimizing risks and conflicts. Preparing for the next generation of space applications requires fostering collaboration at a global level. Satellites and orbital platforms achieve their full potential when international cooperation guides development and use.

Technological standards are essential for ensuring compatibility, safety, and efficiency in space operations. Policies that define design, operational, and safety standards for satellites, AI systems, and orbital infrastructure prevent collisions, data errors, and system failures. Standardization also enables interoperability between international partners, private companies, and public agencies. Satellites operate more effectively when technological guidelines are enforced, ensuring reliability, sustainability, and optimal societal benefit. Preparing for next-generation space applications requires developing and enforcing clear standards that guide innovation responsibly. By codifying best practices, policies ensure that orbital technologies serve humanity efficiently and safely. Standards provide a foundation for coordination, accountability, and long-term global benefit.

Ethical oversight is a fundamental component of policy for space technology. Satellites, orbital systems, and AI-driven platforms carry responsibilities related to privacy, environmental impact, and societal welfare. Policies can require ethical assessments, transparent reporting, and continuous monitoring to ensure compliance with moral principles. By embedding ethics into regulations, governments and international bodies prevent misuse, inequity, and unintended harm. Satellites and space-based infrastructure serve humanity most effectively when ethical considerations are central to governance. Preparing for next-generation applications means ensuring that technological innovation aligns with societal values, environmental sustainability, and global equity. Ethical oversight transforms space policy into a mechanism that guarantees responsible and beneficial use of orbital technology.

Sustainability is a critical focus of space policy, ensuring that satellites and orbital infrastructure do not harm the environment or jeopardize long-term space operations. Policies can mandate debris mitigation, environmentally responsible launch practices, and sustainable materials for spacecraft. By integrating sustainability into regulations, governments encourage the development of technologies that protect both Earth and orbiting ecosystems. Satellites and orbital systems maximize societal benefit when designed and operated with environmental responsibility in mind. Preparing for next-generation space applications requires policies that balance innovation with ecological preservation. Space technology becomes a durable and ethical tool when sustainability principles guide its development, launch, and operations.

Disaster preparedness is another area where policy enhances the benefits of space technology. Satellites provide real-time monitoring of natural hazards such as hurricanes, floods, wildfires, and earthquakes, offering critical data for emergency response. Policies can mandate that satellite data is shared promptly with governments, NGOs, and local communities to optimize preparedness and relief efforts. By integrating space technology into national and international disaster management frameworks, policymakers ensure faster, more effective responses. Satellites serve humanity most effectively when their applications are governed by policies that prioritize safety, equity, and timely access. Preparing for future challenges requires regulatory mechanisms that embed space-based tools into comprehensive disaster strategies. Policy transforms satellite networks into instruments that save lives and protect communities worldwide.

Regulation of the private sector is essential for responsible development and deployment of space technology. Companies operate satellites, orbital infrastructure, and AI platforms, often at scales larger than national space agencies. Policies can establish licensing, operational guidelines, safety standards, and ethical requirements to prevent monopolization, misuse, or environmental harm. Public-private partnerships guided by policy encourage innovation while ensuring accountability and public benefit. Satellites achieve their full societal impact when private sector activity aligns with national and global regulations. Preparing for the next generation of space applications requires adaptive policy frameworks that govern both technological advancement and corporate responsibility. Policy ensures that commercial space efforts contribute positively to global well-being.

Climate initiatives are an important domain where space policy has a significant effect. Satellites provide essential data for tracking greenhouse gas emissions, monitoring deforestation, measuring ocean temperatures, and evaluating ice coverage. Policies can require standardized reporting, global collaboration, and open data access to ensure that climate monitoring informs mitigation and adaptation strategies effectively. Space technology supports sustainable development when policy encourages cooperation between nations, public agencies, and private companies. Preparing for next-generation applications requires integrating space-based climate monitoring into policy frameworks that prioritize planetary health. Satellites become indispensable tools for addressing climate challenges when governance aligns technology with environmental objectives.

International treaties form the foundation for cooperative, safe, and ethical use of space technology. Agreements like the Outer Space Treaty, the Moon Agreement, and other multilateral frameworks guide the peaceful use of satellites, orbital infrastructure, and space resources. Policies informed by treaties ensure that space technology is shared equitably, operated safely, and developed sustainably. Satellites maximize their societal value when deployed under the guidance of international law, standards, and cooperative frameworks. Preparing for the next generation of space applications requires compliance with, and enhancement of, treaties to address emerging technologies and challenges. Space policy anchored in international agreements promotes collaboration, trust, and global benefit.

Holistic policy strategies are essential to maximize the benefits of space technology for society. Integrating regulations for innovation, sustainability, disaster preparedness, climate monitoring, and international cooperation ensures that satellites and orbital systems deliver their full potential. By coordinating cross-sector policies, governments can harmonize technological advancement with ethical standards, public safety, and equitable access. Satellites and orbital infrastructure are most effective when policies provide a comprehensive framework that guides research, development, deployment, and usage. Preparing for next-generation applications requires foresight, collaboration, and a commitment to long-term societal impact. Policy transforms space technology into a tool for global progress, ensuring innovation serves humanity rather than narrow interests.

Governance for emerging technologies ensures that new space systems, AI platforms, and orbital infrastructure operate safely and ethically. Policies must anticipate challenges from autonomous satellites, mega-constellations, and space-based manufacturing. Clear regulations, adaptive oversight, and continuous monitoring maintain reliability and prevent misuse. Satellites achieve their greatest impact when guided by governance that balances innovation, safety, and societal benefit. Preparing for future applications requires policies that evolve alongside technology to address risks and opportunities effectively. Space policy safeguards the responsible development and deployment of cutting-edge innovations.

Equitable space access guarantees that satellites, orbital platforms, and AI-driven applications benefit all nations and communities. Policies can promote affordability, shared infrastructure, and knowledge transfer to prevent disparities in access to space technology. By embedding equity into regulatory frameworks, governments ensure that space innovation contributes to global development rather than reinforcing existing inequalities. Satellites serve humanity most effectively when policies prioritize inclusion, shared resources, and capacity building. Preparing for next-generation applications requires proactive strategies that empower developing nations and underserved populations. Equitable policy ensures that the transformative potential of space technology is widely distributed.

Ethical frameworks are integral to responsible space policy. Satellites and orbital systems have wide-ranging implications for privacy, security, environmental protection, and societal welfare. Policies can require ethical assessments, transparent reporting, and accountability mechanisms to align technological deployment with moral principles. By embedding ethics into governance, space technology can deliver benefits while minimizing risks and harm. Preparing for future applications means ensuring that ethical considerations are central to every stage of research, development, and operations. Satellites maximize societal value when innovation is guided by ethical responsibility, fairness, and global cooperation.

Long-term societal benefits are the ultimate measure of successful space policy. Comprehensive, adaptive, and ethical frameworks ensure that satellites and orbital infrastructure address climate challenges, disaster response, energy management, and sustainable development. Coordinated policies, international cooperation, and regulatory foresight create conditions for continuous innovation while safeguarding equity, safety, and the environment. Preparing for next-generation space applications ensures that orbital technologies improve quality of life and foster resilience for generations to come. Space policy transforms innovation into a force for sustainable, inclusive, and transformative global impact. By establishing thoughtful frameworks today, humanity can harness satellites and space-based systems to build a safer, more prosperous, and equitable future.

Chapter . 40

A Sustainable Earth Through Space Innovation

Illustrating a roadmap where space exploration and technology solve Earth's most pressing problems.

Space innovation holds the potential to transform sustainability efforts on Earth, addressing climate change, resource management, and global development. Satellites, orbital energy platforms, and AI-driven space systems provide critical data and capabilities for monitoring the environment, optimizing energy use, and guiding informed policy decisions. By integrating space technology into sustainability strategies, governments, private companies, and communities can enhance resilience, efficiency, and equity. Preparing for a sustainable future requires a coordinated approach, combining research, technology deployment, and ethical oversight. Satellites become powerful tools for positive change when space innovations are applied to solve real-world problems. Space exploration not only extends human presence into orbit but also creates solutions that improve life on Earth.

Environmental monitoring is one of the most immediate benefits of space innovation. Satellites track deforestation, ice melt, ocean temperatures, air quality, and greenhouse gas emissions in real time. This data enables governments and organizations to implement targeted climate mitigation and adaptation strategies. Remote sensing from orbit provides insights into ecosystems, agriculture, and urban development, ensuring that resources are managed sustainably. By leveraging satellite data, communities can make informed decisions to reduce environmental impact and protect biodiversity. Preparing for a sustainable Earth requires continuous investment in space-based observation systems. Space technology amplifies humanity's ability to understand, protect, and manage the natural world effectively.

Renewable energy applications benefit directly from space technology. Orbital solar power stations and advanced satellite systems can support energy distribution, forecast consumption patterns, and optimize renewable energy grids. Satellites provide accurate solar, wind, and hydrological data to maximize efficiency and reduce reliance on fossil fuels. Policies and technological innovations must ensure equitable energy access, particularly in underserved regions. By integrating space technology with energy infrastructure, nations can reduce carbon emissions, enhance reliability, and improve sustainability. Preparing for a sustainable future requires using space innovation to advance energy solutions that meet both environmental and social objectives. Satellites and orbital platforms can transform renewable energy planning, delivery, and management for a greener Earth.

Disaster resilience is enhanced through space innovation. Satellites and AI systems provide early warning, monitoring, and predictive analytics for hurricanes, floods, earthquakes, and wildfires. This information allows governments, NGOs, and communities to respond proactively, minimizing damage, saving lives, and preserving ecosystems. Coordinated use of space technology in disaster management ensures timely relief and long-term recovery planning. Preparing for a sustainable Earth requires integrating orbital insights into emergency response and climate adaptation strategies. Space innovation becomes a force for resilience when satellites and AI provide actionable intelligence for vulnerable regions. Satellites not only detect crises but also guide solutions that protect people, infrastructure, and the environment.

Global collaboration is key to maximizing space innovation for sustainability. International partnerships enable shared satellite constellations, data exchange, joint research, and coordinated environmental initiatives. By collaborating across borders, nations can leverage technology efficiently, reduce duplication, and ensure that space-based solutions benefit all regions equitably. Preparing for a sustainable future requires fostering global networks that integrate technological expertise, policy alignment, and ethical oversight. Satellites and orbital platforms achieve maximum impact when their use is coordinated, transparent, and focused on collective benefit. Space innovation provides a framework for international cooperation that addresses Earth's most pressing challenges. By working together, humanity can harness orbital technology to secure a sustainable and resilient planet.

Sustainable agriculture is greatly enhanced through space innovation. Satellites provide real-time data on soil health, crop growth, irrigation needs, and pest infestations. By analyzing multispectral and hyperspectral imagery, farmers can optimize planting schedules, reduce water usage, and minimize chemical inputs. This precision agriculture approach increases yields while reducing environmental impact, contributing to food security and ecosystem preservation. Policies and technologies that integrate satellite insights with local farming practices ensure equitable benefits for smallholders and large-scale operations alike. Preparing for a sustainable Earth requires leveraging orbital data to make agriculture more efficient, resilient, and environmentally responsible. Space technology enables smarter farming that supports both productivity and sustainability goals.

Water management is another domain where satellites provide critical solutions. Space-based observation allows monitoring of freshwater resources, river flow, groundwater levels, and reservoir capacity in real time. These insights help governments, utilities, and communities optimize water allocation, prevent shortages, and detect contamination early. Remote sensing enables large-scale watershed management, ensuring that water use is sustainable across agriculture, industry, and domestic consumption. Preparing for a sustainable future requires integrating space technology into water governance systems to balance supply and demand efficiently. Satellites provide actionable intelligence that prevents waste, protects ecosystems, and supports long-term water security.

Space innovation thus becomes a cornerstone of sustainable water resource management.

Urban planning benefits significantly from insights provided by space technology. Satellites track population growth, land use, transportation patterns, and environmental stressors in cities. This information supports sustainable infrastructure development, efficient energy distribution, and climate-resilient urban design. By integrating satellite data into planning processes, municipalities can reduce congestion, lower emissions, and optimize resource allocation. Preparing for a sustainable Earth requires applying space-derived insights to create smart, resilient, and environmentally responsible cities. Satellites and orbital systems provide the tools needed for informed urban decision-making that balances growth with ecological preservation. Space innovation ensures that urbanization occurs in harmony with sustainability objectives.

Materials innovation is also supported by space technology. Research conducted on orbital platforms, such as microgravity experiments and advanced manufacturing in space, leads to stronger, lighter, and more sustainable materials. These materials can be applied in construction, transportation, renewable energy, and manufacturing industries on Earth. By using insights from space-based experiments, engineers can create products that reduce energy consumption, extend lifespan, and minimize environmental impact. Preparing for a sustainable future requires leveraging orbital research to improve the materials that underpin modern society. Satellites and orbital labs provide knowledge that transforms manufacturing practices and contributes to sustainability. Space innovation drives technological breakthroughs that enhance both performance and environmental responsibility.

Resource monitoring is a fundamental application of satellites for sustainability. Satellites track forests, mineral deposits, fisheries, and energy resources, providing data for responsible extraction and conservation. By monitoring ecosystems in real time, governments and organizations can implement policies to prevent overexploitation and preserve biodiversity. Integrating space-based resource monitoring into decision-making ensures that economic development occurs without compromising environmental health. Preparing for a sustainable Earth requires using satellites to manage natural resources efficiently, equitably, and sustainably. Space technology transforms how humanity interacts with the planet, enabling data-driven stewardship. Satellites and orbital systems empower societies to balance growth with environmental preservation.

Energy efficiency is greatly improved through space technology. Satellites provide real-time monitoring of electricity grids, solar farms, and wind energy systems, allowing operators to optimize distribution and reduce waste. Orbital data can forecast energy demand, detect inefficiencies, and guide infrastructure improvements, ensuring that renewable energy resources are used effectively. By integrating space-based insights into energy management, governments and companies can reduce carbon emissions and support sustainable development. Preparing for a sustainable Earth requires leveraging satellites to enhance energy efficiency at both local and global scales. Space innovation ensures that energy systems operate intelligently, minimizing environmental impact while meeting growing demand. Satellites become critical tools for transitioning to a cleaner, more efficient energy future.

Climate adaptation relies heavily on data from space technology. Satellites track changes in temperature, precipitation, sea levels, and extreme weather events, providing the information necessary to design resilient communities and infrastructure. Governments can use orbital insights to develop early warning systems, guide urban planning, and implement adaptive agricultural strategies. Preparing for a sustainable Earth requires integrating climate data into policy, planning, and investment decisions to mitigate risks and protect vulnerable populations. Satellites and orbital systems provide the predictive power needed to respond proactively to climate challenges. Space innovation allows societies to anticipate environmental changes, reducing harm and enhancing resilience.

By using satellites to guide adaptation strategies, humanity can prepare for a rapidly changing planet.

Disaster mitigation is strengthened by space innovation. Satellites detect and monitor hurricanes, earthquakes, floods, wildfires, and other natural hazards, enabling early warnings and informed response strategies. By providing precise data in real time, orbital systems allow governments, NGOs, and communities to allocate resources efficiently, evacuate populations, and reduce property damage. Preparing for a sustainable future requires embedding satellite-based disaster mitigation tools into national and international emergency frameworks. Satellites and space-enabled AI systems transform disaster preparedness from reactive to proactive approaches. Space innovation ensures that societies can respond quickly and effectively to crises, protecting lives and ecosystems. Orbital technology thus becomes an indispensable tool for reducing disaster risk and enhancing resilience.

Renewable resource management benefits from space-based observation and analysis. Satellites monitor forests, fisheries, water bodies, and agricultural lands to guide sustainable extraction, conservation, and utilization practices. By providing continuous, comprehensive data, satellites allow governments and organizations to enforce regulations, detect overuse, and optimize resource allocation. Preparing for a sustainable Earth requires policies and technological systems that integrate orbital monitoring into resource management strategies. Satellites and orbital platforms provide actionable intelligence that ensures natural resources are used responsibly and preserved for future generations. Space innovation enables the global community to balance economic development with environmental sustainability. By using satellites to manage renewable resources, humanity can safeguard ecosystems while meeting societal needs.

International cooperation is essential to maximize the benefits of space innovation for sustainability. Shared satellite constellations, data exchange agreements, and joint research initiatives allow countries to address global challenges collaboratively. Policies that encourage cooperation reduce redundancy, lower costs, and promote equitable access to space technology. Preparing for a sustainable future requires fostering multinational partnerships that integrate expertise, resources, and governance structures. Satellites and orbital systems achieve the greatest impact when countries work together to monitor climate, manage resources, and respond to disasters. Space innovation thus becomes a unifying force, connecting nations in pursuit of a sustainable and resilient planet. Collaborative approaches ensure that the benefits of space technology are shared globally, advancing both human welfare and environmental stewardship.

Technological innovation in transportation benefits greatly from space-derived data. Satellites track traffic patterns, air quality, weather conditions, and fuel consumption, helping cities and countries optimize transportation networks. Space-based GPS and navigation systems provide real-time information that reduces congestion, saves fuel, and lowers emissions. By applying space technology to transportation systems, governments can create more efficient, safer, and sustainable mobility options for both people and goods. Preparing for a sustainable Earth requires the integration of satellite systems into transportation planning, from public transit to autonomous vehicles. Satellites not only improve the efficiency of transportation networks but also play a critical role in reducing their environmental footprint. Space innovation enables smarter, greener, and more sustainable transportation infrastructure.

Sustainable urban development is increasingly reliant on space technology. Satellites provide detailed data on land use, urban heat islands, infrastructure needs, and population growth, allowing cities to plan smarter and more resilient developments. By integrating satellite imagery into urban planning processes, policymakers can design cities that optimize energy use, minimize waste, and incorporate green spaces. Space technology also helps monitor air quality, water availability, and climate resilience, ensuring that urban areas can adapt to environmental changes. Preparing for a sustainable Earth requires space-based solutions to create cities that are not only efficient but also environmentally responsible and socially inclusive. Satellites become essential tools for building smart cities that balance development with sustainability goals. By using space technology, urban planning can embrace sustainability while improving the quality of life for residents.

Healthcare applications also benefit from space innovation, particularly through remote sensing and satellite data. Space-based systems enable the monitoring of disease outbreaks, tracking environmental conditions that impact public health, and providing real-time communication for healthcare delivery in remote areas. Satellites assist in disaster response, providing critical data for medical teams and humanitarian organizations. Space technology also supports telemedicine, offering connectivity to underserved regions where healthcare access is limited. Preparing for a sustainable Earth requires integrating space-enabled health solutions that improve global healthcare access, reduce disparities, and respond rapidly to health crises. Satellites and space systems enable more equitable and efficient healthcare delivery, particularly in regions with limited infrastructure. Space innovation has the potential to revolutionize health management and disease prevention worldwide.

Monitoring biodiversity and environmental health is another area where space technology plays a key role. Satellites track forest cover, coral reef conditions, wildlife migration, and habitat destruction, providing essential data for conservation efforts. Space-based sensors enable large-scale environmental monitoring, guiding policies and actions that protect ecosystems and prevent biodiversity loss. By tracking changes in ecosystems, satellites support the enforcement of environmental regulations and the identification of areas in need of protection. Preparing for a sustainable Earth requires policies that integrate satellite data into biodiversity monitoring and conservation efforts. Space technology provides a powerful tool to safeguard the planet's diverse ecosystems and species. Satellite-enabled monitoring helps prevent the degradation of natural habitats and promotes sustainable coexistence between humanity and the natural world.

Global equity is a central consideration in maximizing the benefits of space technology for sustainability. Space innovation has the potential to bridge gaps between developed and developing countries by providing access to critical data, technology, and resources. Satellites enable low-cost access to information that can enhance agricultural productivity, improve disaster preparedness, manage resources, and monitor environmental changes in underserved regions. By promoting policies that ensure equitable access to space-derived benefits, governments and organizations can foster development that is inclusive and sustainable. Preparing for a sustainable Earth requires international collaboration that ensures all nations, regardless of their economic status, have access to the tools and data needed to address global challenges. Space technology empowers communities by providing them with the knowledge and resources to solve local problems while contributing to global sustainability. Space innovation drives equity by ensuring that the benefits of technology are shared broadly and fairly.

Space technology holds the promise of driving sustainable development on Earth by offering innovative solutions to some of our planet's most pressing challenges. From climate change monitoring to resource management, space-derived insights empower governments, organizations, and individuals to make data-driven decisions that protect our environment, enhance societal welfare, and improve global resilience. Satellites and orbital systems provide unparalleled data that guides informed policy, supports emergency responses, and accelerates progress in key sectors such as agriculture, energy, and healthcare. As the world faces an ever-growing array of environmental and social challenges, space innovation serves as a critical enabler for sustainability, bridging gaps in knowledge and capacity. Preparing for a sustainable Earth requires embracing space technology not just as a tool for exploration, but as a cornerstone of global development. By applying space-derived solutions to real-world problems, we can mitigate climate risks, promote equity, and foster a resilient, prosperous future.

International policy plays a pivotal role in maximizing the potential of space technology for sustainability. Collaborative frameworks and agreements are essential to ensure that the benefits of space innovation are shared equitably across nations and sectors. Policy can guide the responsible use of space technology, ensuring that data is accessible, ethical standards are upheld, and environmental impacts are minimized. By fostering international cooperation, governments can avoid fragmentation and ensure that space exploration remains a global endeavor focused on collective human advancement. Space policy must prioritize sustainable practices, encourage cross-border partnerships, and provide access to satellite data for nations and communities most in need. As space technology continues to evolve, proactive policy frameworks will be essential to safeguard its benefits for future generations.

Looking ahead, the future potential of space innovation for addressing Earth's challenges is vast. Advancements in satellite technology, AI, and orbital infrastructure will continue to shape how we monitor our planet, address global crises, and promote long-term sustainability. Space-based solar power, lunar resource extraction, and advanced climate modeling are just a few of the emerging fields that have the potential to revolutionize our approach to sustainability. The next frontier of space exploration will likely bring unprecedented opportunities for both technological breakthroughs and global cooperation. Preparing for the future requires both the development of new technologies and the implementation of policies that ensure these innovations serve humanity and the planet equitably. Space exploration, once solely focused on venturing beyond Earth, is now a vital tool for solving the most pressing problems we face on our home planet.

The role of space innovation in achieving a sustainable Earth is undeniable. As humanity continues to push the boundaries of space exploration, the benefits we derive from this endeavor must be harnessed to tackle the immense challenges we face on Earth. Space technology, when aligned with global sustainability goals, can become an essential instrument in building a safer, cleaner, and more equitable world. By continuing to invest in and refine space technology, supporting international cooperation, and ensuring that sustainability remains at the heart of space exploration, we can unlock new pathways to a more resilient and prosperous future. The journey to a sustainable Earth through space innovation has only just begun, and the possibilities are limitless.

Epilogue:

As we stand at the threshold of a new era, the benefits of space technology stretch before us, offering possibilities we once only dreamed of. From protecting our environment and improving healthcare to managing resources more sustainably and creating smarter cities, space-based innovations have the potential to solve the very challenges that threaten our future.

Yet, the true measure of this progress will not just be in the technology itself but in how we use it. The future is not determined by the tools we have but by the values we bring to their use. How can we ensure that space technology uplifts every community, that it brings equity and opportunity, and that it serves the health of both our planet and our people?

The answers are in our hands. Space has already provided us with the tools; it is up to us to shape a future that reflects our deepest commitments to sustainability, cooperation, and innovation. We must move forward with wisdom, purpose, and a deep respect for both the wonders and responsibilities that space has entrusted to us.

The journey ahead is an extraordinary one. Together, we can ensure that the legacy we leave behind is one of progress, equity, and a world that thrives—both here on Earth and beyond.

About the Author

Laura Lee is a creative visionary, writer, and digital entrepreneur with a passion for bridging emotional depth with innovative technology. Raised in a small town where stories were the currency of connection, she learned early on the transformative power of authentic expression.

As the founder of a forward-thinking digital design agency, Laura has helped brands discover their voice, craft impactful identities, and build meaningful connections with their audiences. Her work —both in design and writing—echoes themes of authenticity, empowerment, and conscious creation. She believes that creativity is not just about aesthetics—it's about intention, and that true innovation starts with a deep connection to purpose.

Through her books, digital platforms, and consulting, Laura empowers others to tap into their creativity, honor their personal journeys, and bring their unique visions to life with integrity and clarity. Her writing is both deeply personal and universally resonant, reflecting the journey of self-discovery, healing, and growth.

This book is an extension of her mission—a love letter to the process of self-discovery, and a guide to finding peace in a world that constantly asks us to forget who we are. Laura's work is driven by a vision to help others cultivate lives rooted in beauty, authenticity, and a higher sense of purpose, both personally and in their global communities.

www.ingramcontent.com/pod-product-compliance
Lightning Source LLC
LaVergne TN
LVHW052222200926
848481LV00001B/189